T0317733

UMTS Performance Measurement

UMTS Performance Measurement

A Practical Guide to KPIs for the UTRAN Environment

Ralf Kreher

Tektronix MPT Berlin GmbH & Co. KG
Germany

John Wiley & Sons, Ltd

Other Wiley Editorial Offices

John Wiley & Sons Inc., 111 River Street, Hoboken, NJ 07030, USA

Jossey-Bass, 989 Market Street, San Francisco, CA 94103-1741, USA

Wiley-VCH Verlag GmbH, Boschstr. 12, D-69469 Weinheim, Germany

John Wiley & Sons Australia Ltd, 42 McDougall Street, Milton, Queensland 4064, Australia

John Wiley & Sons (Asia) Pte Ltd, 2 Clementi Loop #02-01, Jin Xing Distripark, Singapore 129809

John Wiley & Sons Canada Ltd, 6045 Freemont Blvd, Mississauga, Ontario, Canada L5R 4J3

Wiley also publishes its books in a variety of electronic formats. Some content that appears in print may not be
available in electronic books.

British Library Cataloguing in Publication Data

A catalogue record for this book is available from the British Library

ISBN-13 978-0-470-03249-7 (HB)
ISBN-10 0-470-03249-9 (HB)

Typeset in 10/12 pt Times by Thomson Digital.
Artwork by Brit Kreher, Berlin, Germany.

FSC
Mixed Sources
Product group from well-managed
forests and other controlled sources
Cert no. SGS-COC-2953
www.fsc.org
© 1996 Forest Stewardship Council

Contents

Preface

Having dealt with in-depth analysis of SS#7, GSM and GPRS networks I started to monitor UTRAN interfaces approximately four years ago. Monitoring interfaces means decoding the data captured on the links and analysing how the different data segments and messages are related to each other. In general I wanted to trace all messages belonging to a single call to prove if the network elements and protocol entities involved worked fine or if there had been failures or if any kind of suspicious events had influenced the normal call proceeding or the call's quality of service. Cases showing normal network behaviour have been documented in Kreher and Ruedebusch (*UMTS Signaling*. John Wiley & Sons, Ltd, 2005), which provides examples for technical experts investigating call flows and network procedures.

While still writing the last paragraphs of *UMTS Signaling* it became obvious that the focus of leading UMTS technology experts was changing more and more from the investigation of functional behaviour to the analysis of huge data streams supplied by signalling information and user data/payload. As a result the idea of a second book was already born before the first one was ready to be published. Some major customer projects I have been involved in pushed my ideas and knowledge further into this field. Indeed, if one compares radio-related information in UMTS and GSM radio access network protocols, e.g. the contents of measurement reports sent to the network by mobile stations and base stations, it is obvious that in UMTS much more radio-specific measurements are executed. Reports are sent more frequently and by using more sophisticated methods than in GSM to guarantee the quality of service in UMTS networks.

The radio technology behind UMTS is seen in two different varieties: frequency duplex division (FDD, also known as WCDMA), where uplink and downlink data is transmitted on two different frequency bands; and time division duplex (TDD), where uplink and downlink channels are separated using timeslots. TDD is actually beyond the scope of this book, because it has not been introduced in European and North American networks so far. The Chinese solution of a low chip rate TDD (TD-SCDMA) has not yet been deployed in the field, and although deployment may start during 2006 it will take a while before performance measurement becomes crucial for TD-SCDMA operators. First they have to set emphasis on the execution of functional tests. Nevertheless, many measurement definitions and key performance indicators presented in this book will also be valid in TDD networks apart from mostly radio-related measurements and soft handover analysis, because there is no soft handover in TDD.

Many ideas and definitions in UMTS performance measurement scenarios are not described in international standards. There is a big *grey zone* that covers a wide range of proprietary definitions. An examination of these proprietary requirements written by network equipment manufacturers and network operators was a main impetus to write this book. As a result more than three-quarters of the contents deal with descriptions and definitions that cannot be found in any international standard document. And very often proprietary requirements do not entirely depict all necessary measurement details. It is another aim of this book to close the gap between proprietary and 3GPP performance measurement definitions as well as the gap between the theory of measurements and actual implementation.

Ralf Kreher
Berlin, Germany

Acknowledgements

I would like to take the chance to acknowledge the effort of all who participated directly or indirectly in creating and publishing this book.

First of all a special 'thank you' goes to my spouse Grit who volunteered for initial proofreading of all texts and to my sister Brit who created and formatted all the figures you will find in this book. Also my daughters Alva and Luise must be mentioned who brightened a couple of hard working days with their smiles.

This book would not exist without the ideas, questions and requirements contributed by customers, colleagues, subcontractors and competitors. Besides all others that cannot be personally named I would like to express thanks especially to the following people listed in alphabetical order:

Alessio Biasutto
Roberto Cappon
Ilija Cutura
Norbert Eggert
Kaushik Gohel
Rajasekhar Gopalan
Steffen Hülpüsch
Per Kangru
Spiros Kapoulas
Uwe Keuthe
Jens Künzel
Johnson Liu
Martin McDonald
Andrea Nicchio
Marco Onofri
Jürgen Placht
Christian Rust
Alexander Seifarth
Christopher Semturs
Alberto Visetti
Mike Wiedemann

A very important input for this book was the data collected in laboratories and live networks all around the world by Tektronix staff and subcontractors. Thanks go to Daniele Rampazzo, Bhal Vyas, Than Aye, Bernd Wessling and Oliver Schwarz who provided most of the recordings. Analysis of this data would have been impossible without the work of the engineers who participated in creating an amazing software called the *Tektronix UTRAN Network and Service Analyzer.*

In addition thanks go to former Tektronix MPT director of marketing Othmar Kyas and present director of marketing Toni Piwonka-Cole who supported the idea of writing this book and approved usage of Tektronix material in the contents.

Last but not least I also would like to express my thanks to the team at John Wiley & Sons, Ltd, especially Mark Hammond, Jennifer Beal, Tessa Hanford and Sarah Hinton, for their strong support.

1

Basics of Performance Measurement in UMTS Terrestrial Radio Access Network (UTRAN)

Performance measurement represents a new stage of monitoring data. In the past monitoring networks meant decoding messages and filtering which messages belong to the same call. Single calls were analysed and failures were often only found by chance. Performance measurement is an effective means of scanning the whole network at any time and systematically searching for errors, bottlenecks and suspicious behaviour.

Performance measurement procedures and appropriate equipment have already been introduced in GSM and 2.5G GSM/GPRS radio access networks as well as in core networks, however, compared to the performance measurement requirements of UTRAN those legacy requirements were quite simple and it was relatively easy to collect the necessary protocol data as well as to compute and aggregate appropriate measurement results.

Nowadays even Technical Standard 3GPP 32.403 (*Telecommunication Management. Performance Management (PM). Performance Measurements – UMTS and Combined UMTS/GSM*) contains only a minimum set of requirements that is not much more than the tip of an iceberg. The definitions and recommendations of 3GPP explained in this chapter do not cover a wide enough range of possible performance measurement procedures, some descriptions are not even good enough to base a software implementation, and in some cases they lead to completely wrong measurement results. To put it in a nutshell it looks like the specification of performance measurement requirements for UTRAN is still in an early phase. This first part of the book will explain what is already defined by 3GPP, which additional requirements are of interest and which prerequisites and conditions always have to be kept in mind, because they have an impact on many measurement results even if they are not especially highlighted.

By the way, in the author's humble opinion, *the biggest error in performance measurement is the copy and paste error.* This results from copying requirements instead of developing concepts and ideas of one's own. As a result this book will also not contain ready-to-use performance measurement definitions, but rather discuss different ideas and

UMTS Performance Measurement: A Practical Guide to KPIs for the UTRAN Environment   Ralf Kreher
© 2006 Ralf Kreher

offer possible solutions for a number of problems without claiming to cover all possibilities and having the only solutions.

1.1 GENERAL IDEAS OF PERFORMANCE MEASUREMENT

Performance measurement is fairly unique. There are many parameters and events that can be measured and many measurements that can be correlated to each other. The number of permutations is infinite. Hence, the question is: what is the right choice?

There is no general answer except perhaps the following: A network operator will define business targets based on economical key performance indicators (KPIs). These business targets provide the guidance to define network optimisation targets. And from network optimisation targets technical KPI targets can be derived, which describe an aspired behaviour of the network. Based on this, step by step, services are offered by operators. On a very common level these are e.g. speech calls and packet calls. These services will be optimised and detected errors will be eliminated. All in all it is correct to say that the purpose of performance measurement is to troubleshoot and optimise the network (see Figure 1.1).

However, whatever network operators do, it is up to the subscriber to finally evaluate if a network has been optimised in a way that meets customers' expectations. A rising churn rate (i.e. number of subscribers cancelling a contract and setting up a new one with a competitor operator) is an indicator that there might also be something wrong in the technical field.

Fortunately there is very good news for all analysts and market experts who care about churn rates: it is very difficult to calculate a real churn rate. This is because most subscribers in mobile networks today are prepaid subscribers, and since many prepaid subscribers are

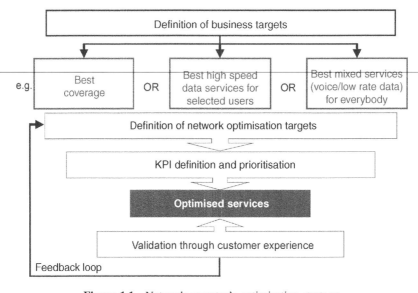

Figure 1.1   Network operator's optimisation strategy

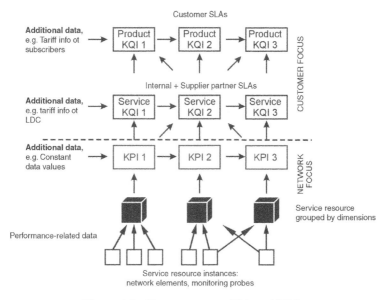

Figure 1.2   How to compute KPIs and KQIs

people who temporarily stay abroad, and based on the fact that prepaid tariffs are often significantly cheaper than roaming tariffs, such subscribers become temporary customers, so to speak. Once they go back to their home countries their prepaid accounts remain active until their contracts expire. Therefore not every expired contract is a churn. The actual number of churns is expected to be much less, but how much less? Additional information is necessary to find out about this.

The fact that additional information is necessary to compute non-technical key performance indicators based on measurement results (in this case based on a counter that counts the number of cancelled and expired contracts) also applies to the computation of technical KPIs and key quality indicators (KQIs). See Figure 1.2.

The general concept of these indicators is that network elements and probes, which are used as service resource instances, are placed at certain nodes of the network infrastructure to pick up performance-related data, e.g. cumulative counters of protocol events. In constant time intervals or in near real time this performance-related data is transferred to higher level service assurance and performance management systems. A typical example for such a solution is Vallent Corporation's WatchMark® software that is fed with performance data sent by radio network controllers (RNCs), mobile switching centres (MSCs) and GPRS support nodes (GSNs). For this purpose, e.g. an RNC writes the values of its predefined performance counters into a predefined XML report form every 15 minutes. This XML report file is sent via a so-called northbound interface that complies with the Tele Management Forum (TMF) CORBA specification to WatchMark® or any other higher level network management system. Additional data such as traffic and tariff models are provided by other sources and finally a complete solution for business and service management is presented.

As pointed out in www.watchmark.com the overall solution:

... provides benefits across a service provider's entire customer base including pre-paid, post-paid and enterprise customers:

- Service quality management provides an end-to-end visibility of service quality on the network to ensure that each service (e.g. MMS, WiFi, iMode, SMS and GPRS etc.) is functioning correctly for each user on the network.
- Internal and 3rd Party service level agreements (SLAs) allow Service Providers to test, evaluate and monitor service levels within the organization to ensure that optimum service quality is delivered to customers.
- Corporate SLAs enable Service Providers to establish specific agreements with their corporate customers where they undertake to deliver customized end-to-end levels of service quality.

However, there is one major problem with this concept: network elements that feed higher level network management systems with data are basically designed to switch connections. It is not the primary job of an RNC to measure and report performance-related data. The most critical part of mobile networks is the radio interface, and the UTRAN controlled by RNCs is an excellent place to collect data giving an overview of radio interface quality considering that drive tests that can do the same job are expensive (at least it is necessary to pay two people per day and a car for a single drive test campaign). Secondly, performance data measured during drive tests cannot be reported frequently and directly to higher layer network management systems. Therefore a great deal of important performance measurement data that could be of high value for service quality management is simply not available. This triggers the need for a new generation of measurement equipment that is able to capture terabytes of data from UTRAN interfaces, performs highly sophisticated filtering and correlation processes, stores key performance data results in databases and is able to display, export and import these measurement results using standard components and procedures.

Before starting to discuss the architecture of such systems it is beneficial to have a look at some definitions.

1.1.1 WHAT IS A KPI?

Key performance indicators can be found everywhere, not just in telecommunications. A KPI does not need to deal with only technical things. There are dozens of economical KPIs that can be seen every day, for example the Dow Jones Index and exchanges rates. The turnover of a company should not be called a KPI, because it is just a counter value, however, the gross margin is a KPI. Hence, what makes the difference between performance-related data and a KPI is the fact that a KPI is computed using a formula.

There are different kinds of input for a KPI formula: cumulative counter values, constant values, timer values seem to be the most important ones. Also KPI values that have been already computed are often seen in new KPI formulas.

Most KPI formulas are simple. The difficulties are usually not in the formula itself, but e.g. in the way that data is first filtered and then collected. This shall be demonstrated by using a simple example. Imagine a KPI called *NBAP Success Rate*. It indicates how many

NBAP (Node B application part) procedures have been completed successfully and how many have failed.

NBAP is a protocol used for communication between Node B (the UMTS base station) and its CRNC (controlling radio network controller). To compute a *NBAP Success Rate* a formula needs to be defined. In 3GPP 25.433 standard for Node B Application Part (NBAP) protocol or in technical books dealing with the explanation of UMTS signalling procedures (e.g. Kreher and Ruedebusch, 2005) it is described that in NBAP there are only three kinds of messages: Initiating Message, Successful Outcome and Unsuccessful Outcome (see Figure 1.3).

Following this a *NBAP Success Rate* could be defined as shown in Equation (1.1):

$$\textbf{NBAP Success Rate} = \frac{\sum NBAP\ Successful\ Outcome}{\sum NBAP\ Initiating\ Message} \times 100\% \qquad (1.1)$$

This looks good, but will lead to incorrect measurement results, because an important fact is not considered. There are two different classes of NBAP messages. In class 1 NBAP procedures the Initiating Message is answered with a Successful Outcome or Unsuccessful Outcome message, which is known in common protocol theory as acknowledged or connection-oriented data transfer. Class 2 NBAP procedures are unacknowledged or connectionless. This means only an Initiating Message is sent, but no answer is expected from the peer entity.

Since most NBAP messages monitored on the Iub interface belong to unacknowledged class 2 procedures (this is especially true for all NBAP common/dedicated measurement reports) the *NBAP Success Rate* computed using the above defined formula could show a value of less than 10%, which is caused by a major KPI definition/implementation error.

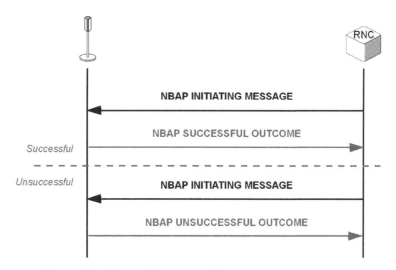

Figure 1.3   Successful/Unsuccessful NBAP call flow procedure

Knowing the difference between NBAP class 1 and class 2 procedures a filter criteria needs to be defined that could be expressed as follows:

$$\textbf{NBAP Class 1 Success Rate} = \frac{\sum NBAP\ Successful\ Outcome}{\sum NBAP\ Class\ 1\ Initiating\ Message} \times 100\% \quad (1.2)$$

An exact definition is usually not expressed in formulas, but more often by fully explaining in writing the KPI definition. A couple of examples can be found in Chapter 2 of this book. The lesson learnt from the *NBAP Success Rate* example is that one cannot compare KPIs based on their names alone. KPIs even cannot be compared based on their formulas. When KPIs are compared it is necessary to know the exact definition, especially the filter criteria used to select input and – as explained in next chapter – the aggregation levels and parameter correlations.

Never trust the apparently endless lists of names of supported KPIs that can be found in marketing documents of network and measurement equipment manufacturers. Often these lists consist of simple event counters. Therefore, it must be kept in mind that additional data is always necessary as well as simple counter values to compute meaningful KPIs and KQIs.

1.1.2 KPI AGGREGATION LEVELS AND CORRELATIONS

KPIs can be correlated to each other or related to elements in the network topology. The correlation to a certain part of the network topology is often called the aggregation level.

Imagine a throughput measurement. The data for this measurement can be collected for instance on the Iub interface, but can then be aggregated on the cell level, which means that the measurement values are related to a certain cell. This is meaningful because several cells share the same Iub interface and in the case of softer handover they also share the same data stream transported in the same Iub physical transport bearer that is described by AAL2 SVC address (VPI/VCI/CID). So it may happen that a single data stream on the Iub interface is transmitted using two radio links in two or three different cells. If the previously mentioned throughput measurement is used to get an impression of the load in the cell it is absolutely correct to correlate the single measurement result with all cells involved in this softer handover situation.

To demonstrate the correlation between mobile network KPIs an example of car KPIs shall be used (see Figure 1.4). The instruments of a car cockpit show the most important KPIs for the driver while driving. Other performance-relevant data can be read in the manual, e.g. volume of the fuel tank.

The first KPI is the speed, computed by the distance driven and the period of time taken. Another one is the maximum driving distance, which depends on the maximum volume of fuel in the tank. Maybe the car has an integrated computer that delivers more sophisticated KPIs, such as fuel usage depending on current speed, and the more fuel needed to drive a certain distance influences the maximum driving distance. In other words, there is a correlation between fuel usage and the maximum driving distance.

Regarding mobile telecommunication networks like UTRAN similar questions are raised. A standard question is: How many calls can one UTRA cell serve?

Network equipment manufacturers' fact sheets give an average number used for traffic planning processes, e.g. 120 voice calls (AMR 12.2 kbps). There are more or less calls if

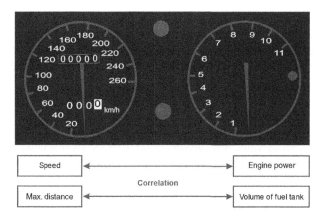

Figure 1.4   Correlation between car KPIs

different services such as 384 kbps data calls or different AMR codecs with lower data transmission rates are used. The capacity of a cell depends on the type of active services and the conditions on the radio interface, especially on the level of interference. Hence, it makes sense to correlate interference measurements with the number of active calls shown per service. This combination of RF measurements requires sophisticated KPI definitions and measurement applications. The first step could start with the following approach: Count the number of active connections per cell and the number of services running on those active connections in the cell.

Before continuing with this example it is necessary to explain the frame conditions of this measurement, looking at where these counters can be pegged under which conditions and how data can be filtered to display counter subsets per cell and per service.

1.1.3 BASIC APPROACH TO CAPTURE AND FILTER PERFORMANCE-RELATED DATA IN UTRAN

The scope of this book is UTRAN performance measurement. Within UTRAN four interfaces exist where performance-related data can be captured: the Iub interface between Node Bs and RNC; the Iur interface between different RNCs; the IuCS interface between RNCs and the CS core network domain; and the IuPS interface between RNCs and the PS core network domain. For each interface a specific protocol stack is necessary to decode all layers of captured data as explained in detail in Section 1.2, which deals with the functions and architecture of performance measurement equipment. Usually this equipment is able to automatically detect to which specific interfaces a probe is connected and which protocol stacks are necessary to decode captured data. If necessary it can also detect on which particular channel data is transmitted. This especially refers to dedicated and common transport channels on the Iub interface. In addition, it can be assumed that the same equipment also provides a function that is commonly known as *call trace*, which allows for the automatic detection and filtering of all messages and data packets belonging to a particular connection between a single UE and the network. For a detailed overview of all interfaces, channels and call procedures it is recommended to read the appropriate chapters in Kreher and Ruedebusch (2005). From a performance measurement expert's perspective it is expected that these functions are provided and work as required to decode and aggregate performance-related data. Nevertheless, in this chapter a few basic network procedures need to be explained, that apply to all scenarios, because they may be relevant for any call or at any time during an active connection.

Our approach is as already defined in the previous section. Count all connections in a cell and provide a set of sub-counters that is able to distinguish which services are used during these connections. From a subscriber's perspective this scenario is simple. They switch on their mobile phones, set up calls, walk around or drive by car (which could result in a couple of mobility management procedures) and finish their calls whenever they want.

Now from a network operator's perspective it is necessary to find out in which cell the calls are active and identify the type of service related to each particular call. It sounds easy, but due to the specific nature of the UTRAN procedure it is indeed a fairly complicated analysis.

When the term 'service' is used in the context of performance measurement this usually applies to end-user services such as voice calls, data calls and – if available in network and if the UE is capable – video-telephony calls. All kinds of supplementary services such as conference calls or multi-party connections are seen as special cases of the above categories and are not analysed in detail. However, when looking at data calls the type of service can also be determined from the TCP/IP application layer, e.g. file transfer (FTP) or web browsing (HTTP). These specific services are beyond the scope of this basic approach for two reasons. Firstly they require a more complex correlation of measurement data, secondly it makes no sense to define a TCP/IP analysis at cell level, because even the smallest email or website is segmented by the RLC into a number of different transport blocks and theoretically each transport block set can be transmitted using a different cell.

There is another well-known service from GSM, which is also available in UMTS. This service is called short message service (SMS). A short message is not sent using a dedicated traffic channel, it is sent piggybacked on signalling messages. Plain signalling is also necessary to register a mobile phone to the network after being switched on. There is no payload transmitted between subscriber and network, but nevertheless signalling is essential and for this reason another service type called 'signalling' will be defined in addition to 'voice', 'data' and 'video-telephony' in this basic approach.

Now the question is how to distinguish the four different services by monitoring protocol messages.

A CS call set up always starts with a Call Control Setup message as specified in 3GPP 24.008. The 'decision maker' that distinguishes between voice calls and video-telephony calls is the value of the bearer capability information element within this Setup message. If the bearer capability information element shows the value 'unrestricted digital info' the call is a video-telephony call. Another indicator is the signalling access protocol I.440/450 and rate adaptation following H.223 & H.245 mentioned in the same message. See Figure 1.5.

It is difficult to explain what a bearer is. Maybe the following definition is the best one: A bearer is a temporary channel used to transport a data stream (user or network data) with a defined quality of service. (All definitions in this book are given by the author using his own words. Standard definitions may be more exact, but are often not very understandable.)

This is true for both GSM and UMTS, but in UMTS the bearer concept covers all possible data streams in each part and layer of the network while in ISDN/GSM it is only used to

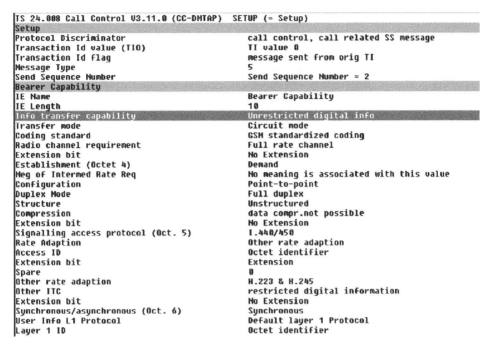

```
TS 24.008 Call Control V3.11.0 (CC-DMTAP)   SETUP (= Setup)
Setup
Protocol Discriminator                      call control, call related SS message
Transaction Id value (TIO)                  TI value 0
Transaction Id flag                         message sent from orig TI
Message Type                                5
Send Sequence Number                        Send Sequence Number = 2
Bearer Capability
IE Name                                     Bearer Capability
IE Length                                   10
Info transfer capability                    Unrestricted digital info
Transfer mode                               Circuit mode
Coding standard                             GSM standardized coding
Radio channel requirement                   Full rate channel
Extension bit                               No Extension
Establishment (Octet 4)                     Demand
Neg of Intermed Rate Req                    No meaning is associated with this value
Configuration                               Point-to-point
Duplex Mode                                 Full duplex
Structure                                   Unstructured
Compression                                 data compr.not possible
Extension bit                               No Extension
Signalling access protocol (Oct. 5)         I.440/450
Rate Adaption                               Other rate adaption
Access ID                                   Octet identifier
Extension bit                               Extension
Spare                                       0
Other rate adaption                         H.223 & H.245
Other ITC                                   restricted digital information
Extension bit                               No Extension
Synchronous/asynchronous (Oct. 6)           Synchronous
User Info L1 Protocol                        Default layer 1 Protocol
Layer 1 ID                                  Octet identifier
```

Figure 1.5   Call Control Setup message for video-telephony call

define the characteristics of traffic channels between subscribers. A service from the point of view of UTRAN is always bound to a certain type of (radio) bearer and hence, analysing characteristics of UMTS bearer services is another possible definition of 'call type' and is completely different from the approach given in this chapter which is based on NAS signalling analysis.

Looking back to the specific signalling used between the UE and the CS core network domain it emerges that in contrast to video-telephony calls voice calls have the bearer capability value 'speech' in the Call Control Setup message. A PS connection (data call) always starts with a Service Request message. This Service Request indicates that there is data (IP payload) to be transmitted, but it should be noted that this definition might not always fit to the user's perspective of an active PS call.

Imagine a subscriber starting a mobile web-browsing application. For this purpose a PDP context is established between the UE and the SGSN and a traffic channel, which is called the radio access bearer (RAB) is provided. Now a website is downloaded and the user starts to read its contents. This may take a while. Besides the user may switch to another application while keeping the web-browser open. This is not a problem in fixed data networks. IP data is only transmitted when necessary, if there is no data transfer no network resources of the fixed line are occupied. That does not apply to UTRAN. Here dedicated resources (these are the codes used to identify channels on the radio interface) need to be provided for each RAB. And those resources are limited. That is the reason why the network needs to identify which resources are really used. All other resources are released to prevent shortage and guarantee subscriber satisfaction. This leads to a situation that a PDP Context that is bound to the open web-browsing application remains active in the UE and SGSN while a RAB is released if the network detects that no data is transmitted for a certain time. Based on this a PS connection in UTRAN is defined, whereas an active PS RAB and RAB assignment is always triggered by a Session Management Service Request message.

RABs are also set up for CS connections, but for conversational calls they are active as long as a call is active. Indeed, there are several ways to count the number of active connections, which means that there are different protocol messages from different protocol layers. The advantages of the method described in this chapter are:

1. Non-access stratum (NAS) signalling messages can be found on both Iub and Iu interfaces.
2. NAS messages contain information elements that allow direct identification of the call type. In the case of e.g. an RANAP RAB Assignment Request it can only be guessed from the UL/DL maximum bit rate and traffic class mentioned in this message which call type is related to the RAB. This requires additional mapping tables running in the background of the performance measurement application (note that this is an alternative option).
3. Setup and Service Request messages may contain user identifiers that allow further filter options (e.g. count all active connections per cell, per call type, per UE) and are helpful for troubleshooting.

To complete the call type definition, 'signalling' constitutes all call flows between the UE and core network domains that do not contain Setup or Service Request messages. It is also necessary to define another category that is usually called 'Multi-RAB' and describes a UE that has at least two active connections (RABs) simultaneously. Multi-RAB calls can be a combination of CS and PS services for one UE, but multiple PS RABs are also possible, for instance if PS streaming video requires the set up of a secondary PDP context that triggers the establishment of a second PS RAB for the same UE. This second RAB provides a different traffic class (= different delay sensitivity) and different maximum bit rates. An example for such a kind of Multi-RAB PS+PS would be a GPRS session management message Activate Secondary PDP Context Request. Figure 1.6 shows the different filter options.

Protocol events used to determine the call type cannot immediately be used to count the number of active connections, because they only describe connection attempts. Therefore, it is necessary to check if the attempted connection has been set up successfully. This can be done on the RRC, RANAP or NAS layer. On the Iub interface the RRC Radio Bearer Setup Complete message indicates that a traffic channel has been established successfully. Following this the RANAP RAB Assignment Response is sent on the Iu interface while the NAS layer indicates that the connection between A-party and B-party has been established. For PS calls the session management Service Accept and PDP Context Activation Accept messages could be used as additional indicators for a successful connection. It should be noted that in the case of video telephony calls via the CS domain in-band signalling is also necessary to really get the service running. This in-band signalling is transmitted using the radio (access) bearer and the example proves that there are different perspectives of user and network and it clarifies the need to have different KPIs for those different perspectives.

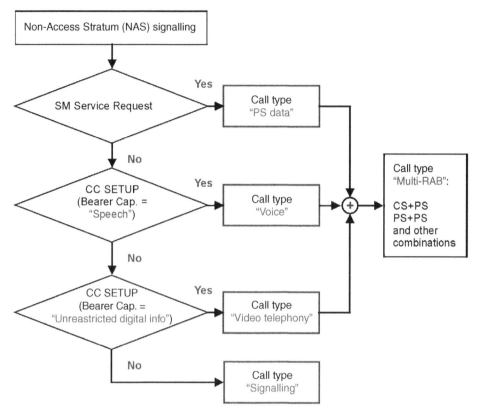

Figure 1.6   Filter options for call type

Having defined the necessary set of counters to determine the number of active connections, it is important to find out which calls are active in which cell.

When counting the messages on Iu interfaces there is actually no chance of finding a relationship between call and cell. There is only one identifier that allows a very vague estimation of the cell by using the service area code (SAC), which is part of the service area identity (SAI). A single SAC represents a single cell according to the definition in the numbering plans of most UMTS network operators. Normally it is the cell in which the UE is located when the call is set up. This cell can easily be detected on the Iub interface, because the first message of each call (RRC Connection Request or RRC Cell Update – on the basis that Paging messages only trigger calls) is always sent on the random access channel (RACH) and there is always one and only one RACH assigned to each cell.

A problem is caused by the following. When the first message of a call is monitored on the RACH there is no identifier that indicates which cell this RACH is related to. Certainly the CRNC knows which RACH belongs to which cell, but this correlation can only be monitored when the RACH is installed immediately after the Cell Setup procedure. It remains unknown as long as the channel is not deleted and re-established or reset. Hence, cell ID related to RACH must be found by monitoring the system and using an indirect algorithm. This algorithm is based on the fact that one or more forward access channels (FACHs) belong to the same cell as the RACH and the messages sent on an FACH are called RRC Connection Setup or Cell Update Complete. These messages contain a cell identifier and have a 1 : 1 relationship with RRC Connection Request or RRC Cell Update Request messages previously monitored on the RACH. Using this principle the cell identity related to a certain RACH can be investigated and the results are stored in a special database of the performance measurement system. This database is often named the 'topology module'. It contains necessary information to correlate protocol events with network elements and their identifiers.

Due to the fact that the phone is a mobile it moves and may change the cell, which causes another problem. The footprint of UMTS cells is on average 10 times smaller than the footprint of GSM cells and therefore cell changes, called 'handovers' in UMTS, occur much more often. In addition, most currently deployed UMTS networks in Europe and North America work in FDD mode. FDD stands for frequency division duplex and means that the uplink and downlink channels use different frequency bands. This applies for instance for the already mentioned RACH (uplink) and FACH (downlink).

A special kind of handover, the soft handover, is possible between FDD cells working on the same frequency bands for uplink and downlink radio transmission. A hard handover as known from GSM – also defined for UMTS – is characterised by the fact that first a call is deleted in the old cell (also named 'source cell') during handover and interrupted for a very short time frame of max. 200 ms before it is continued in a new cell (called 'target cell'). However, in the case of a soft handover there is no interruption, instead the user equipment (UE) is connected to several cells simultaneously. The UE has a radio link to each cell involved in a soft handover situation. A bundle of all radio links belonging to a single UE is called its active set.

Preventing call interruptions caused by hard handover procedures is not the main reason for a soft handover. Soft handovers are interesting because they allow the transmitted power of UEs to be kept as low as possible, even if UEs are located at cell borders as shown in Figure 1.7.

Figure 1.7 show that two radio links (= cells) are part of the active set. By the definition of 3GPP, depending on UE capabilities and on the capabilities of the UTRAN, the UE can use up to six cells simultaneously for a certain period of connection via radio interface while it has just one active connection with the network on the NAS level.

Under these circumstances there is a new aspect to the original task of counting the number of active connections per cell, which is user mobility. There are two different levels of mobility management in the network, mobility management on the NAS level, which is mainly executed using location update and inter-MSC handover procedures in the CS domain and routing area update procedures in the PS domain. Both were already introduced in GSM/GPRS networks. In UTRAN a new level of mobility management can be found, which is controlled by the RNC. This mobility management is executed using RRC signalling messages and enables cell changes as well as radio link additions and deletions in the case of soft handover situations.

If the analysis target is to find out how many connections are active in each cell it is necessary to track the mobility of each single UE in the network, because the number of connections per cell depends on the location of the UEs. Due to this fact a relatively simple counter already requires highly sophisticated measurement applications and it is not enough just to count some protocol events.

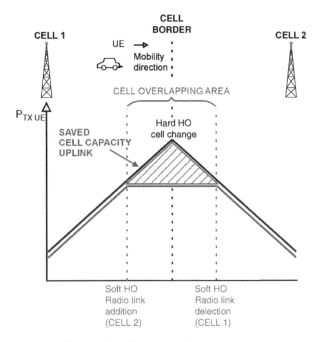

Figure 1.7   The soft handover advantage

The next sections will give an overview of the requirements for UTRAN performance measurement which have been defined by 3GPP and of additional measurement procedures that have not been defined by standard organisations, but are also necessary to deliver useful performance measurement results.

1.1.4 PERFORMANCE MEASUREMENT DEFINITIONS OF 3GPP

Anyone searching for KPI definitions in 3GPP specifications will be disappointed. 3GPP has not defined any key performance indicators. However, there are definitions of performance-related data that is either transmitted between network elements using measurement report messages or data measured in network elements and provided for analysis in higher level network management systems. Basically those definitions cover two major groups of performance-related data: radio-related measurements and protocol event counters.

Also only a fairly general framework document exists that describes performance measurement tasks from a high level perspective. This document is 3GPP 32.401 'Concepts and Requirements for Performance Measurement'. It defines all tasks for measurement job administration, e.g. the start and stop of measurement procedures and how measurement jobs are initialised, modified and terminated. The second part deals with measurement type definitions and measurement result generation. It contains e.g. descriptions of how gauges are measured (e.g. max/min/mean values of response time measurements and when cumulative counters need to be reset). In addition, there are some interesting statements about granularity and accuracy, e.g. a success event will always belong to the same sampling period as the attempt event to which it is bound. All in all, this standard document gives a good overview on how to store and transport performance data that is already available, but is says nothing about how to get such data or about how to compute and analyse it.

Measurement definition can be found in 3GPP 32.403 'Performance Measurement – UMTS and Combined UMTS/GSM'. The title sounds like a perfect compendium, but actually it is just a list of protocol event counter definitions. It is highly recommended to look for the latest available document version, because in some older versions (e.g. v4.5.0) even the protocol events are not always well defined, as the following example shows. It defines for instance a single counter that counts NBAP and RNSAP radio link addition events, but in the two protocols – despite the procedures having the same name – they have totally different functions. Newer versions like v5.8.0 have corrected this error and define separate counters for NBAP and RNSAP radio link addition.

All in all the main limitations of 3GPP 32.403 remain the same, no matter which specific document version is used:

- Only counters for higher layer protocol layers are defined – no events are defined that indicate that problems might appear on transport layers, for instance if AAL2 SVCs cannot be set up or are dropped.
- From an RNC perspective only the RRC, NBAP and RNSAP dealing with connection set up and handover are taken into account. There is no definition of ALCAP events, neither are counters for other interesting events defined (e.g. NBAP radio link failure indication).
- There is no multi-interface perspective, e.g. it is not possible to distinguish between NBAP radio link setup for intra-RNC soft handover and inter-RNC soft handover (triggered by the RNSAP radio link setup or radio link addition procedures).
- There is no multi-layer perspective, e.g. it is not possible to have combined soft handover analysis based on NBAP, ALCAP and RRC events that interact to perform this mobility management procedure.
- And finally there is no relation between defined protocol events and network services as they can be found in service level agreements between network operator and subscriber.

Also there is another problem with precise counter definitions. Usually protocol events are defined on a very general (which means not specific) level. It is one thing to mention that hard handover procedures can be executed using RRC physical channel reconfiguration, RRC transport channel reconfiguration, RRC radio bearer reconfiguration, RRC radio bearer establishment or RRC radio bearer deletion, however, there is no information about which message is used in which case and which information element would distinguish between intra-frequency and inter-frequency hard handover (as shown in Figure 1.8). Such specific information is also missing in 3GPP 25.331 *Radio Resource Control*.

Some of those limits might be covered by definitions found in 3GPP 52.402 *Telecommunication Management; Performance Management (PM); Performance Measurements – GSM*. This standard document is well structured and covers more aspects than 32.403, but it does not contain any UTRAN-relevant definitions. It could be seen as a combination of contents of 32.401 and 32.403 with emphasis on GERAN. Obviously it would be correct to call 3GPP 52.402 a sister document of 3GPP 32.403.

A much better situation is found when looking for definitions of radio-related measurements. Here the specifications are written in a detailed and complex way although it does

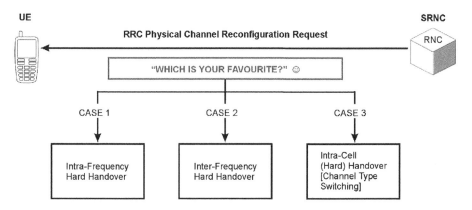

Figure 1.8   Same protocol message used for different handovers procedures

take a lot of knowledge to understand what is measured, how measurement results are computed, transmitted and finally presented.

It starts with 3GPP 25.215 *Physical Layer – Measurements (FDD)*. This standard document gives a comprehensive overview of measurement abilities on UE and UTRAN. It is very useful to understand which parameters are measured in and reported by UE (using RRC measurement reports) and cell/Node B (using NBAP measurement reports). For each measurement parameter a clear definition is given including exceptions and limitations. 3GPP 25.216 is a comparable document of 3GPP 25.215 for UTRAN TDD mode but details of radio quality measurement in TDD mode are beyond the scope of this book.

The most important radio measurement abilities defined in 3GPP 25.215 are:

From Node B (NBAP measurement reports):

- Common measurements – related to a cell, not to a single connection:

  - **Received total wideband power** (RTWP) = total UTRA uplink frequency noise on cell antenna

  - **Transmitted carrier power** = total downlink Tx power of cell antenna

  - **Preambles of PRACH** = number of connection request attempts on radio interface necessary until network reacts by sending either positive or negative confirmation. In the case of positive confirmation UE is allowed to send an RRC Connection Request as the first higher layer message.

- Dedicated measurements – related to a single connection or single radio link, but not related to a single cell:

  - **SIR** (signal-to-interference ratio) = ratio between measured uplink RSCP of a single UE's signal and the interference code power (ISCP) after spreading of received signals

- **SIR error** – measured difference between uplink SIR target set by SRNC and measured SIR

- **Transmitted code power** = transmitted power for one downlink dedicated physical channel (DPCH) sent by a single cell/antenna

- **Round trip time** on Uu interface = estimation based on time difference between RLC frames and appropriate RLC Acknowledgement (RLC AM)

From UE (RRC measurement reports). All those measurements are related to a single radio link (= single cell) used for a single connection, because they all depend on the location of the UE related to a single cell at the time of measurement:

- **Chip energy over noise (Ec/N0)** = downlink equivalent of SIR, but based on measurement of common pilot channel (CPICH)
- **Received signal code power (RSCP)** – Rx level of downlink dedicated physical channel (DPCH) on UE antenna
- **UTRA received signal strength indicator (RSSI)** – total UTRA downlink frequency noise on UE antenna, downlink equivalent to RTWP.
- **Event-IDs** – used to report predefined measurement events measured downlink frequency.

There are other measurement abilities defined in the same 3GPP standard document, but they have not been implemented yet in UTRAN, which can be understood as a sign of lower priority.

All the above-mentioned radio-related measurements will be explained in more detail in Section 2.2.

While 3GPP 25.215 defines the measurement parameters there is another interesting specification named 3GPP 25.133 *Requirements for Support of Radio Resource Management (FDD).* It explains how the measurement results of 3GPP 25.215 parameters are reported, which reporting ranges are defined and how these measurement results are encoded in signalling messages (measurement reports). There are also some formula definitions that describe how radio-specific measurement values are computed in UE or Node B, but these formulas are not KPI definitions. Additionally, measurement requirements for all UTRAN mobility management procedures such as cell (re-)selection and handovers and RRC connection control are described including timing and signalling characteristics related to these procedures.

These few standard documents contain what is defined for UTRAN performance measurement in 3GPP standards and it is always a good idea to remember these standard documents when talking about KPI definitions.

1.1.5 USER EXPERIENCE VS. 3GPP PERFORMANCE MEASUREMENT DEFINITIONS

Remembering the network operator's optimisation strategy (Figure 1.1) it is a fact that the subscriber's experience – often called *user perceived quality of service* or *user quality of experience* – is a key factor of business success. Hence, it is important for network operators to measure this quality of experience (QoE). From the previous section one question arises:

Are 3GPP performance measurement definitions of an acceptable standard to measure the user quality of experience?

This leads to another question: What do users recognise when they use network services?

Imagine a subscriber using a UMTS mobile phone. All he wants is to switch it on, make a voice call, make a data call and switch it off. If everything works fine the subscriber's impression is positive. If there are problems they will most likely fall into one of the following five categories:

- Subscribers are not able to register to the network.
- Subscribers are not able to set up calls.
- The calls made are dropped before the calling or the called party hangs up.
- Poor information transmission speed is measured especially for data calls, e.g. by file transfer software.
- The quality of transmitted information is bad, which especially has an impact on conversation calls (voice, video-telephony).

While subscribers are not interested in the details of problem analysis, technicians and engineers working for network operators need to find the root causes and measure the occurrence of problems.

1.1.5.1 Problems with Registration and Call Setup

For registration and call setup procedures it is first necessary that the UE has an active RRC connection with its serving RNC (SRNC). This connection either needs to be set up or it needs to be pushed into RRC states CELL_DCH or CELL_FACH, because only in those states is it possible to exchange NAS messages between UE and network. Also NAS messages are necessary to register and to set up calls.

Following this there is a number of possibilities why the UE is not able to perform registration or call setup procedure:

- RRC Connection cannot be set up because network does not answer UE's request or connection setup request is rejected by the network due to blocking.
- Once signalling radio bearers (the dedicated control channels (DCCHs) that carry RRC messages) are established it is possible that they are dropped, e.g. due to failures on radio interface, handover problems on RRC mobility management level or other reasons.
- Although the RRC connection is working properly problems with establishment of Iu signalling connections using RANAP might turn up.
- And finally problems on the NAS layer may occur that lead to exceptions in registration and call setup procedures. In part 2 of Kreher and Ruedebusch (2005) a troubleshooting example has been given in which a Location Update procedure (a CS registration) is rejected by the MSC due to 'network failure' and root cause analysis has proven that RNC has failed to execute a security mode request sent by CS domain. This example may stand as a typical failure belonging to this group.

All in all there are several dozen single-interface single-layer procedures that could block successful registration or call setup procedures. Most of these single-layer

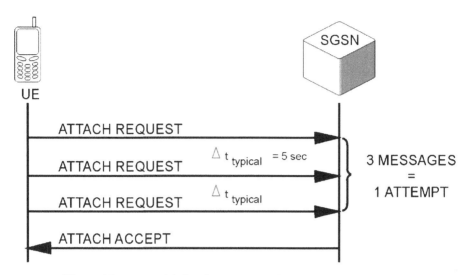

Figure 1.9 Successful GPRS Attach Procedure with multiple attempts

procedures are analysed if performance measurement counters defined in 3GPP 32.403 are implemented in network elements (RNC, MSC, SGSN). However, usually these counter values do not reflect the customer's experience, because highly sophisticated filtering and correlation functions are not implemented in statistics functions of network elements.

Figure 1.9 shows a typical example of a GPRS Attach (PS registration) procedure that is finally successful, however, three successive attempts (Attach Request messages) are necessary to get a successful answer from the network. Such a scenario is seen fairly often in live networks, especially if there is a long distance and hence a long response time between the SGSN and HLR, which often happens if subscribers are roaming in foreign networks far away from their home country.

In this case the user might not recognise a problem at all. Especially since he knows that due to the roaming situation the registration procedure may take longer than in the home Public Mobile Network (PLMN). However, a KPI based on simple counter definitions found in 3GPP 32.403 without special filtering/processing would indicate registration problems, because it would show Attach Success Rate = 33.33%. The KPI formula of such a GPRS Attach Success Rate looks as follows:

$$\text{GPRS Attach Success Rate } (\%) = \frac{\sum GPRS\ Attach\ Accept}{\sum GPRS\ Attach\ Request} \times 100 \qquad (1.3)$$

This does not reflect the subscriber's experience and proves the need of implementing sophisticated filtering/call trace analysis processes in performance measurement software while the formula remains correct and does not need to be changed.

1.1.5.2 Dropped Calls

Subscribers recognise that calls are dropped, that is to say an ongoing connection (defined as successfully established RAB in Section 1.1.3) is terminated for abnormal reasons. Most likely those reasons are found on the radio interface, for instance bad coverage or unsuccessful handovers. There is a signalling message that indicates such problems on Iub. It is the NBAP Radio Link Failure Indication containing the cause value 'synchronization failure'.

In 3GPP 32.403 a counter for this protocol event is not defined. Most likely because 3GPP authors have in mind that it is sufficient to analyse cause values found in RANAP Iu-Release Request or RAB-Release Request messages that trigger the RANAP Iu-Release or single RAB Release procedure. Usually a RANAP cause value like 'radio connection with UE lost' indicates that a particular release request was triggered by an NBAP Radio Link Failure Indication, but not every NBAP Radio Link Failure Indication indicates a dropped call – especially if the user's experience is taken into account. This is true because if a problem on a single radio link of an active set occurs other radio links of the same active set will still continue to transmit data. However, in such cases a bit error rate (BER) and block error rate (BLER) of the connection may arise. For this reason BER and BLER are often seen as indicators of a user's quality of experience. However, even if a radio connection was totally dropped in UMTS it can be re-established. The procedure used for this purpose is called RRC re-establishment, but it does not only include re-establishment of RRC signalling connection – all previously used dedicated transport channels (DCHs) for radio bearers can also be re-established as shown in Figure 1.10.

The problems with the call presented in Figure 1.10 start with an attempted soft handover radio link addition. This is indicated by event-ID *e1a* ('a primary CPICH entered reporting range') sent by the UE using an RRC Measurement Report message. After receiving this measurement report the SRNC tries to install the necessary resources in target cell/target Node B, but Node B indicates that there are not enough user plane resources to set up a second radio link for the active set. Looking at the cause value seen in the NBAP Radio Link Setup Failure message (Unsuccessful Outcome) it is hard to evaluate which resources exactly have not been available, e.g. was there a congestion in transport network or was the code assigned for transmission of dedicated information via the radio interface already in use? Now the SRNC reacts in a way that is seen quite often in similar trouble scenarios. Instead of looking for an alternative solution to guarantee radio link quality of the call all

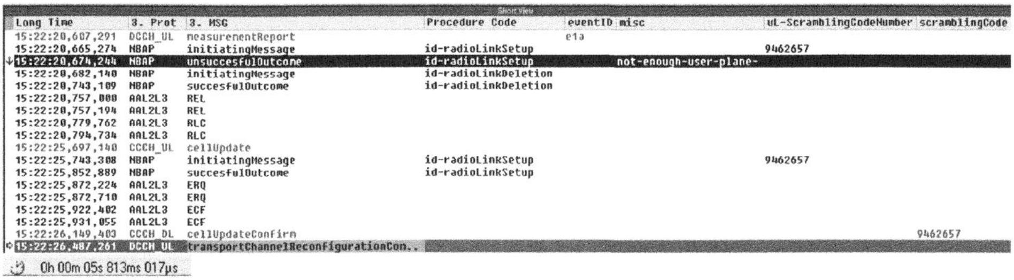

Long Time	3. Prot	3. MSG	Procedure Code	eventID	misc	uL-ScramblingCodeNumber	scramblingCode
15:22:20,607,291	DCCH_UL	measurementReport		e1a			
15:22:20,665,274	NBAP	initiatingMessage	id-radioLinkSetup			9462657	
15:22:20,674,244	NBAP	unsuccesfulOutcome	id-radioLinkSetup		not-enough-user-plane-		
15:22:20,682,140	NBAP	initiatingMessage	id-radioLinkDeletion				
15:22:20,743,109	NBAP	succesfulOutcome	id-radioLinkDeletion				
15:22:20,757,000	AAL2L3	REL					
15:22:20,757,194	AAL2L3	REL					
15:22:20,779,762	AAL2L3	RLC					
15:22:20,794,734	AAL2L3	RLC					
15:22:25,697,140	CCCH_UL	cellUpdate					
15:22:25,743,308	NBAP	initiatingMessage	id-radioLinkSetup			9462657	
15:22:25,852,889	NBAP	succesfulOutcome	id-radioLinkSetup				
15:22:25,872,224	AAL2L3	ERQ					
15:22:25,872,710	AAL2L3	ERQ					
15:22:25,922,402	AAL2L3	ECF					
15:22:25,931,055	AAL2L3	ECF					
15:22:26,149,403	CCCH_DL	cellUpdateConfirm					9462657
15:22:26,487,261	DCCH_UL	transportChannelReconfigurationCon..					

0h 00m 05s 813ms 017µs

Figure 1.10 RRC Re-establishment following Radio Link Failure

Table 1.1 Subscriber's experience of interrupted call related to type of service

Type of service	Subscriber's perceived QoS
Web browsing	Normal (as experienced in fixed network)
File download	Not good, but acceptable as long as there is no critical delay recognised by user
Voice/Video-telephony	Critical impact on user-perceived QoS. Six-second interruption is too much compared with interruption caused by hard handover (approx. 200 ms). If the problem occurs too often user satisfaction will be significantly influenced

radio links are terminated – without informing UE about this radio link deletion (it could have been done using RRC signalling connection). What happens from the UE's perspective is that dedicated radio links are suddenly gone, the mobile phone finds itself alone in the field, falls into CELL_FACH state and sends RRC Cell Update (cause = 'radio link failure') to the best available cell using the RACH of this cell.

The reselected cell forwards this Cell Update message to its controlling RNC (CRNC). By chance in this case it is the same RNC that used to be the serving RNC of the abnormally terminated RRC connection. Hence, this RNC establishes the radio link again using the RRC Cell Update Complete message and after approximately 6 seconds the previously dropped connection becomes active again. Final statement: the call is not really dropped, but only interrupted for approximately 6 seconds.

Now the question is: Which experience of provided network service does the subscriber have if a call is interrupted for 6 seconds? The answer depends on the type of service that has been used (see Table 1.1).

Once again this example proves how important it is to correlate technical measurement data with service information and common sense for human behaviour. It will also be highlighted that although a particular subscriber's perceived quality of service may be deteriorated by the described behaviour of the RNC the programmers of RNC software have chosen a very efficient way to handle such problems. Here it must be kept in mind that it is the RNC's main task to handle thousands of calls simultaneously. To look for a specific trouble-shooting solution for a particular call would block CPU and memory resources of RNC hardware, which can rather be used to handle a few more additional calls that do not struggle with difficulties. Hence, it is correct to say that the risk that this particular call becomes a dropped call is the price that is paid to increase the overall capacity of the RNC. And to increase the overall switching capacity of the RNC means to increase the overall capacity of the network and thus the network operator's profit.

1.1.5.3 Poor Transmission Speed

The performance parameter used to measure transmission speed is called *throughput*. Throughput can be measured on different protocol levels. In the UTRAN environment the

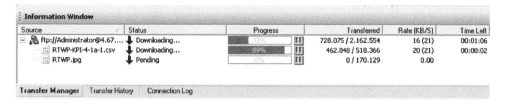

Figure 1.11   Throughput measurement in FTP client software

following throughput measurements are useful:

- RLC throughput to have a traffic measurement that reflects uplink/downlink traffic load on a cell. RLC throughput includes all kinds of user plane/control plane traffic transmitted via the radio interface, no matter if identical information is simultaneously transmitted in any neighbour cells due to soft handover situations.
- Transport channel throughput is the data transmission speed measured on a single transport channel (usually the dedicated transport channel (DCH)) that can be compared with the theoretical possible transport channel throughput derived from used transport format set definitions.
- User perceived throughput is the throughput of RLC payload excluding payload information that is retransmitted due to RLC AM. Secondary traffic due to soft handover situations transmitted on the radio interface need to be subtracted as well.
- Application throughput is measured on the TCP/IP application layer (ISO-OSI Layer 7). This is a throughput of user applications like file transfer and web browsing – to name just two out of approximately 50 possible application protocols. Often subscribers have good tools on their PC to measure application throughput. Figure 1.11 shows an example of how an FTP client shows progress of file transfer and the measured transfer rate in kilobytes per second (it is unclear what the values in parentheses represent). The crucial point is that application throughput measured in the network must be comparable with measurement results available for subscribers.

There are a number of factors that have an enormous impact on throughput, especially radio link quality and propagation conditions. Interference can deteriorate the radio link quality. The required data transmission speed for a certain service is directly related to the coverage of a cell. The basic rule is: the higher the requested data transmission rate of service (transport channel throughput) the less is the coverage provided by the cell under unchanged propagation conditions.

An example shall be given to explain this latter sentence. Imagine a base station that is 30 metres above ground level and a UE used in a average height of 1.5 metres (above ground level). Both endpoints of the radio connection are located in a city centre environment. Now using the same UE in the same cell under the same conditions we set up two connections. Connection #1 is a 12.2 kbps AMR speech call while connection #2 is a 144 kbps data call. How does the type of service (expressed as data transmission speed = throughput) influence the maximum cell range (d)?

Using the COST Hata propagation model with a clutter correction term for the city centre environment it can be calculated that the cell range of speech calls is approximately

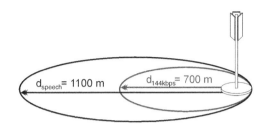

Figure 1.12   Cell range (d) vs. transmission speed of service

1100 metres while the cell range of the same cell for the 144 kbps data service is limited to approximately 700 metres (see Figure 1.12). To put it in a nutshell, UE needs to be closer to the cell antenna to guarantee a higher data transmission rate.

What can be seen in Figure 1.12 is also the reason why UTRAN cells are usually much smaller than GSM cells. Under which conditions can the promised peak data transmission rate of 14.4 Mbps for high speed packet downlink access (HSDPA) be reached?

The 14.4 Mbps, which is usually presented by marketing people and journalists, is theoretically possible with a coding rate of 4/4, 16 quadrature amplitude modulation (QAM) and all available spreading codes (15 in total) bundled for a single connection. Realistic scenarios as introduced e.g. by US consultant Peter Rysavy (www.rysavy.com) in a white paper written for 3G Americas (www.3gamericas.org) mention an average throughput rate (application throughput?) 'for file downloads using HSDPA between 500 and 1100 kbps' and during 2005 the author of this book heard rumours that the optimisation target of some network operators for HSDPA is to guarantee a minimum downlink data throughput of 384 kbps in hot-spot areas (shopping centres, airports etc.) for all HSDPA subscribers – but on which throughput level? These are questions that will have a significant impact on statements regarding throughput measurement results.

Certainly throughput measured for PS services do show quite dramatic changes within a single call, because it is the nature of PS calls that data is only transmitted when necessary and often it is transmitted in peaks. A typical PS call scenario could look like this:

- A user starts web browsing.
- A huge file can be downloaded from a web site. User starts to download.
- While the download is ongoing the user uses different other applications and does not pay attention to the file transfer application that has finally finished the file download.
- After a period of inactivity the user starts web browsing again with short interruptions, e.g. to read the web content, until the web browser application is closed.
- Network problems detected in the last phase of connection are not registered by the user, because of the analysis given in Table 1.1.

An idealised throughput measurement related to this single connection could look as shown in Figure 1.13.

The arrows represent signalling events related to changes in throughput. It is assumed that the HS-DSCH can be used for file download. During this single call there are three

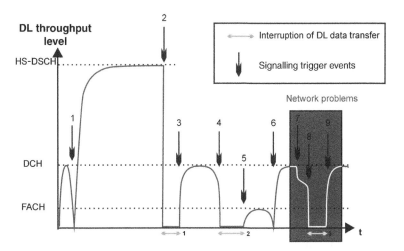

Figure 1.13   Idealised throughput measurement for PS call

interruptions of data transfer. Interruption numbers 1 and 2 are caused by the behaviour of the user, number 3 is caused by radio link failure shown in Figure 1.10. If there is not much data to be transferred SRNC orders the UE to use common transport channels RACH (for uplink) and FACH (for downlink) for data transfer. Since many UEs share the same RACH/FACH at the same time data transmission rates are significantly lower. The dotted lines in the figure mark the maximum theoretical possible throughput associated with the used transport channel. (For DCH there are actually several possible levels depending on the used spreading factor, but this will be discussed in more detail in the next section.)

During the call the following signalling triggers (represented by numbered arrows) occurred:

1. Physical channel reconfiguration: DCH to HS-DSCH after requested file transfer (download).
2. End of download, short interruption (reconfiguration to CELL_PCH or IDLE mode possible).
3. Reconfiguration to CELL_DCH, user continues web browsing.
4. New interruption of web browsing due to user behaviour, again reconfiguration of transport resources.
5. Web browsing with low traffic – only common transport channels are used to transport payload.
6. Reconfiguration to CELL_DCH because of rising DL data amount.
7. Loss of DL throughput due to changing conditions on radio interface (e.g. if user moves) – soft handover attempt.
8. Radio link deletion following unsuccessful soft handover – contact with UE lost.
9. Successful re-establishment of dedicated resources after UE performed Cell Update procedure (cause = 'radio link failure')

Following this there are three major factors that have an impact on the measured PS transport channel throughput:

- user behaviour;
- radio resources (channels, codes) assigned by network;
- network failures.

User behaviour cannot be measured, it can only be estimated based on measurement events. For the other two factors it is necessary to define a number of useful measurements and KPIs.

However, there are not only PS data services running in UTRAN. Another throughput measurement example shows the results for five subsequent video-telephony calls made in a live UMTS network. Transmission conditions in the cell have been fairly ideal. There was no uplink interference, because the test UE was the only UE in the cell and there was no downlink interference. Also fast moving of the UE can be excluded, because no car was involved in the test scenario, however, it was possible to walk around while speaking and watching a live picture of B-party in the UE display.

The throughput measurement results of those five calls made subsequently within 15 minutes are shown in Figure 1.14. Call number 1 and number 2 have reached the maximum possible throughput level at the beginning, but for some reason they have not been able to stay within this threshold. Call number 3 and number 5 show the optimum graph for throughput measured on the DCH, but call number 4 is a mess. Here in the first third of the time interval of this call an interruption of the data stream is observed, maybe caused by radio link failure as described before. However, it cannot be confirmed as long as there is no correlation of radio quality measurements and signalling events with throughput measurement results.

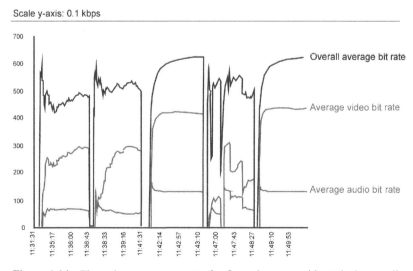

Figure 1.14 Throughput measurements for five subsequent video-telephony calls

This example proves that throughput measurement only detects symptoms of possible network problems, but without correlation to other measurements – discussed in Chapter 2 – no root cause analysis is possible.

1.1.5.4 Corrupted Data

3GPP has been or is aware of the possibility that transmitted user data can be corrupted, especially when transmitted on the radio interface. 3GPP 26.091 explains the methods of error concealment for corrupted AMR speech blocks: 'Normal decoding of lost speech frames would result in very unpleasant noise effects. In order to improve the subjective quality, lost speech frames will be substituted with either a repetition or an extrapolation of the previous good speech frame(s). This substitution is done so that it gradually will decrease the output level, resulting in silence at the output.' In other words: if speech information is corrupted the subscriber will either hear again what he/she has already heard 20 ms before (20 ms is the time interval for transmission of two subsequent voice frames) or in the worst case the subscriber will hear nothing.

In AMR frames we find two different information elements that indicate corrupted speech information (see Figure 1.15). First we have the frame quality indicator that is a single bit in the AMR header. If this bit is set to 1 the expected speech quality in the AMR core frame is good, otherwise it was already bad when the frame was constructed, which is before transmission on radio interface. Possible root causes for high frame quality error rates might be found in encoder entities of UE or MSC/MGW, but especially in the case of traffic coming from network it is possible that speech data received from the B-Party of the call may already have been of bad quality, e.g. due to excessive compression algorithms used by long distance carriers. Or if the B-party is another UMTS subscriber root cause might be a high uplink block error rate (UL BLER) on radio transmission. Whenever bad speech quality is detected by the frame quality indicator, error concealment must be executed.

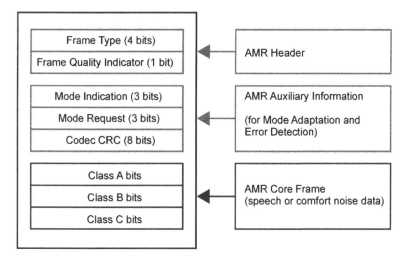

Figure 1.15   AMR generic frame structure (Source: 3GPP 26.101)

Besides the frame quality indicator there is also a codec CRC field consisting of 8 bits. A CRC check is performed for Class A bits of the core frame only, because Class A speech information is most sensitive to errors. By using CRC, data that is transmitted via radio interface is protected. The receiving AMR entity in the UE (after downlink transmission) or the MSC/MGW (after UL data transmission) are able to detect radio transmission errors. If a CRC error is detected 3GPP 26.101 defines that the erroneous AMR frame 'should not be decoded without applying appropriate error concealment'.

Based on the results of CRC checks in the receiving AMR entity a frame error rate (FER) is defined as follows.

$$\text{Frame Error Rate (FER)} = \frac{\sum AMR\ CRC\ Errors}{\sum Received\ AMR\ Frames\ with\ Codec\ CRC} \times 100\% \quad (1.4)$$

Harri Holma and Antti Toskala have outlined that the FER must be below 1% to guarantee sufficient speech quality using AMR (see Holma and Toskala, 2004). The same rule applies to block error rates measured after radio interface transmission, because it must be assumed that a RLC transport block containing currupted speech information on AMR layer is handled like an AMR CRC error.

Unfortunately, quality indicators similar to those defined for AMR are missing for video-telephony. One reason for this is that video-telephony standards have not been defined by 3GPP. A video-telephony call via the CS core network domain is seen by UTRAN as an unspecified symmetric-bidirectional radio access bearer with an uplink/downlink maximum bit rate of 64 kbps (remember bearer capability information contents in the Setup message shown in Figure 1.5). In other words: for UTRAN a video-telephony RAB at the beginning is just an empty channel with a theoretical transmission capacity of 64 kbps in uplink and downlink. Then additional in-band signalling is necessary to start a call between two video-telephones using this previously empty RAB. Following this in-band signalling two logical channels (one for audio, the other one carrying video information bits) are established inside the RAB, but for UTRAN these are just plain data streams, which are transmitted using RLC transparent mode. Hence, there is no error detection/correction executed by UTRAN and also not by any network element in UMTS core network domains. The problem is that video information is very sensitive to transmission errors and in addition used video codices are sensitive as well. Here again BLER measurement results can help to measure the user perceived quality of this service. An alternative way is to reassemble voice and video information captured on UTRAN interfaces to analyse the contents, but this method is only applicable for single call analysis.

Most UMTS handsets are equipped with an MPEG-4 video-telephony codec. MPEG-4 adapts its compression algorithms to enable transmission bandwidth. This bandwidth is equivalent to transport channel throughput. Comparing throughput measurement results presented in Figure 1.13 for call number 1 with the reassembled picture sequence of this call (Figure 1.16), it emerges that resolution of the picture deteriorates if throughput goes down. Some pictures only seem to consist of big 8×8 pixel blocks with sharp edges; therefore the whole picture looks like a chessboard. These effects are caused by a higher compression rate due to lower throughput.

In addition, transmission errors on the radio interface extinguish information of single pixels or pixel blocks. These blocks are missed in subsequent pictures and shown either as

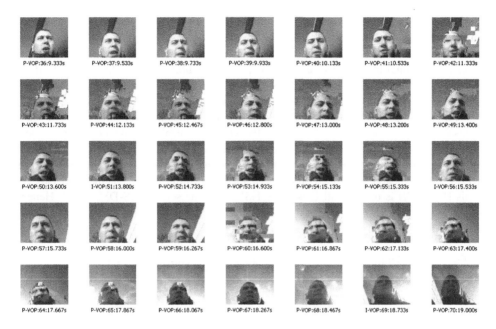

P-VOP:36:9.333s	P-VOP:37:9.533s	P-VOP:38:9.733s	P-VOP:39:9.933s	P-VOP:40:10.133s	P-VOP:41:10.533s	P-VOP:42:11.333s
P-VOP:43:11.733s	P-VOP:44:12.133s	P-VOP:45:12.467s	P-VOP:46:12.800s	P-VOP:47:13.000s	P-VOP:48:13.200s	P-VOP:49:13.400s
P-VOP:50:13.600s	I-VOP:51:13.800s	P-VOP:52:14.733s	P-VOP:53:14.933s	P-VOP:54:15.133s	P-VOP:55:15.333s	I-VOP:56:15.533s
P-VOP:57:15.733s	P-VOP:58:16.000s	P-VOP:59:16.267s	P-VOP:60:16.600s	P-VOP:61:16.867s	P-VOP:62:17.133s	P-VOP:63:17.400s
P-VOP:64:17.667s	P-VOP:65:17.867s	P-VOP:66:18.067s	P-VOP:67:18.267s	P-VOP:68:18.467s	I-VOP:69:18.733s	P-VOP:70:19.000s

Figure 1.16   Sequence of reassembled pictures of video-telephony call #1

white spots or they still show contents of previous pictures that do not fit into the current one (e.g. in the picture with index *P-VOP: 60:16.600s*).

It is known that other video codices like H.263 would fit much better to the requirements of mobile video-telephony, but an error concealment within UTRAN cannot be introduced by changing the used codec. Therefore user perceived quality of service for CS video-telephony is always close to the critical threshold and for this reason some network operators have already decided not to launch or to offer this service anymore.

1.1.6 BASICS OF PS CALL ANALYSIS IN UTRAN

The discussion of throughput measurement in the previous section has shown that it is necessary to have a closer look at the analysis of PS connections in UTRAN since they are more complex and more difficult to understand than simple speech calls. The basic rule for all PS services is that radio resources assigned to these connections are often changed or – as this is called in protocol descriptions – reconfigured. This is necessary because data rates of PS connections change frequently depending on the nature of TCP/IP applications and user behaviour. It is important that subscribers have the feeling of being permanently connected, but there is no need to have a dedicated connection established in UTRAN if no or only small amounts of data need to be transmitted with long time intervals. For instance an email client like Microsoft Outlook may be open permanently and the subscriber wants to be able to send/receive mails at any time. However, data only needs to be transmitted if an email is actually sent or received. Depending on the temporarily necessary bandwidth needed for data transfer different states of PS connections have been defined by 3GPP as well as processes to adapt the provided bandwidth according to actual needs at any time in the call.

The fastest data transmissions are possible using high speed downlink packet access (HSDPA) and – in the future – high speed uplink packet access (HSUPA). This requires the best conditions on the radio interface (pico cells, minimal path loss between sender and receiver and minimal interference) as well as special capabilities of UE and UTRAN. Both parties need to support new modulation techniques like 16 QAM and transport format sets that are different from those used on dedicated channels.

Dedicated transport channels (DCHs) are standard channels for all kinds of data transmission between UE and UTRAN. In the case of voice calls the radio bearer will always be mapped onto DCHs, and will be carried by a set of dedicated physical channels. Dedicated in a sense that those channels are exclusively assigned to a single UE while in HSDPA the radio bearer is mapped onto a high speed downlink shared channel (HS-DSCH) and this HS-DSCH is used by all UEs using HSDPA in the serving HS-DSCH cell. Hence, additional signalling is necessary on the radio interface to inform these UEs which packets sent on HS-DSCH belong to which mobile phone.

Dedicated physical channels are identified by channelisation codes that are also called spreading codes. Scrambling codes provide additional identification of the sender in the uplink or downlink direction.

Channelisation codes and scrambling codes are the radio resources that the UTRAN deals with. Radio resources are assigned by radio network controllers (RNCs).

Unfortunately, the number of available codes is limited and based on requirements of code allocation (e.g. orthogonality) not every possible code can be used. Used channelisation codes also depend on the required data transmission rate. An AMR 12.2 kbps voice channel will work with a spreading factor (SF) of 128 while a 128 kbps PS connection needs SF = 16. However, theoretically only 16 UEs can use a connection with SF = 16 simultaneously in one cell. Indeed, the real number is usually lower due to the necessary orthogonality of codes. The smallest possible spreading factor in UTRAN is SF = 4. Theoretically four UEs can use a channel with SF = 4 at same time in one cell.

Higher data transmission rates require a smaller spreading factor of channelisation codes. The smaller the spreading factor is the fewer codes are available.

Now imagine what happens if spreading factors that allow high data transmission rates are already used and another user wants to set up a PS call in same cell. There are two options. Either the cell is blocked, because it cannot serve the high-speed data transmission request of the subscriber or the UE must accept a lower data transmission rate due to the fact that only codes with higher spreading factors are available.

Due to this shortage of available codes a highly flexible assignment of these critical radio resources has been defined by 3GPP and assigned spreading factors as well as channel mapping options can be changed at any moment of the call.

If only very small amounts of data need to be transmitted there are no dedicated resources assigned at all, because for this purpose, common transport channels RACH (uplink) and FACH (downlink) are used to transport IP payload in addition to the signalling information they usually carry.

If the PDP context remains active, but it is detected that in active connection for a certain time no data is transferred at all, it is also possible to sent the UE back to IDLE mode or to CELL_PCH or URA_PCH mode. In all three modes there is no radio bearer/radio access bearer assigned to the active PDP context – purely for resource usage optimisationreasons.

The admission control function of an RNC frequently measures the data volume sent in DL to the UE and requests traffic volume measurement reports of the UE. Depending on the

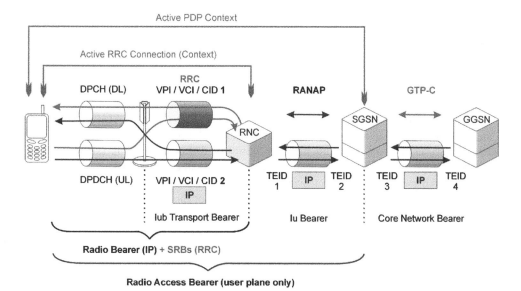

Figure 1.17   Active PDP context – UE in CELL_DCH state

amount of user data temporarily stored in the RLC buffer of the RNC or UE, the SRNC decides to change the spreading factor of the physical channels, change the used transport channel and/or change the assigned RRC state. An overview will be given of how this happens in several stages.

Usually when a PDP context is set up and RAB is assigned for the first time, the UE is ordered by the serving RNC to enter the RRC CELL_DCH state. Dedicated resources (spreading codes, scrambling codes) are assigned by the SRNC and hence, dedicated physical channels are established on the radio interface. Those channels are used for transmission of both IP payload and RRC signalling as shown in the Figure 1.17. RRC signalling may include exchange of NAS messages between the UE and SGSN.

The spreading factor of the radio bearer depends on the expected uplink/downlink IP throughput. The maximum expected data transfer rate can be found in the RANAP RAB Assignment Request message that triggers the set up of the radio bearer and Iu bearer. The Iu bearer is a GTP tunnel for transmission of IP payload on the IuPS interface between SRNC and SGSN.

Activation of the PDP context results also in the establishment of another GTP tunnel on the Gn interface between SGSN and GGSN. In contrast to IuPS where tunnel management is a task of RANAP, the GTP-C (GPRS tunnelling protocol – control) on the Gn interface is responsible for context (= tunnel) activation, modification and deletion. According to the UMTS bearer model described in 3GPP 23.107 *Quality of Service (QoS) Concept and Architecture*, the GTP tunnel on Gn is called the core network bearer.

On the radio interface radio bearers are always transmitted in combination with a set of signalling radio bearers (SRBs). Those SRBs are the dedicated control channels (DCCHs) necessary to transmit RRC and NAS signalling messages.

In RRC Connected Mode there is a state machine in each RRC entity of the connection (on the UE as well as on the SRNC side) that is keeping track of all changes related to this single RRC connection. RRC entities store information about the current RRC state as well as information about used security algorithms and keys or active channel mapping options. All the details known about an active RRC connection are called the RRC context. In a similar way a PDP context contains all the details about an active PS data connection, e.g. between a web browser or email client software and a server on the web.

Figure 1.17 also shows that RRC signalling and IP payload are transported in different bidirectional Iub physical transport bearers, each realised by a single AAL2 switched virtual connection (SVC) and identified by a combination of VPI/VCI/CID (ATM AAL2 SVC address). Node B acts as a kind of multiplexer/de-multiplexer that combines/splits the Iub traffic. It ensures that on the radio interface uplink and downlink traffic (signalling and payload) are transported in separate uplink/downlink physical channels, which in the case of FDD are working on different frequencies.

Two possible key events may cause a reconfiguration of the radio bearer while the PDP context remains unchanged and active. Either the UE sends an RRC measurement report with event 4B (ASN.1 source code: *e4b*) or a timer in the SGSN expires, which controls the activities of the downlink data flow. The measurement report in the uplink direction signals that the RLC buffer of the UE has been below a certain threshold for a certain period of time (defined by time-to-trigger parameter) while timer expiry on the SRNC side happens if there are no IP packets sent or received for a certain time, typically 10 seconds.

If the trigger event occurs the radio bearer will be reconfigured in the following way. Dedicated resources of the connection will be released and a new mapping of the logical channel onto transport channels is established or activated if it has already been defined. Now common transport channels RACH and FACH are used for both exchange of RRC signalling and transmission of IP payload as seen in Figure 1.18. The transport resources of RACH/FACH are shared by several UEs, which use these channels to exchange RRC

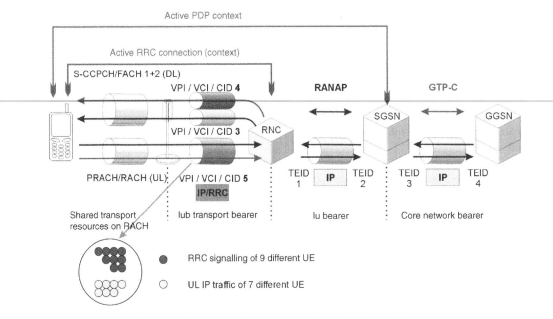

Figure 1.18 Active PDP context – UE in CELL_FACH state

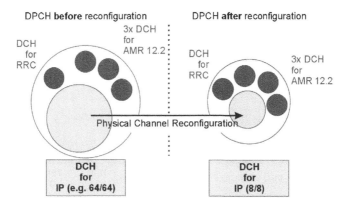

Figure 1.19   Reconfiguration of Multi-RAB connection due to low IP traffic volume

signalling and IP payload in CELL_FACH state, too. Hence, the possible data throughput in this state is very limited. While RACH offers shared transport resources for RRC and IP on the same UL transport channel there are often at least two FACH per cell, the first one for DL RRC, the second one for DL IP payload. These two FACHs are typically mapped onto the same secondary common control physical channel (S-CCPCH – on the radio interface), but use a different Iub transport bearer (VPI/VCI/CID).

This procedure and its reverse process (dedicated resources are set up again after an appropriate RRC measurement report with event ID = 4A or if the SRNC detects a rising RLC buffer level for downlink IP traffic) is also known as (transport) channel type switching.

In the case of a so-called multi-RAB connection (CS and PS connections active simultaneously) the same trigger events that often cause a reconfiguration from CELL_DCH to CELL_FACH state trigger a different kind of physical channel reconfiguration. Due to the active voice call UE cannot leave the CELL_DCH state, but there is also no need to keep the originally assigned resources for a PS connection with e.g. 64 kbps in uplink and downlink. Therefore, after occurrence of the trigger event the RNC performs reconfiguration to adapt used resources for PS connection. Typically the DCH for IP is reconfigured to carry 8 kbps in uplink and downlink. Due to the fact that the overall bandwidth of the multi-RAB connection is not fully required, a new (higher) spreading factor is assigned to the dedicated physical channels. The change of the spreading factor is represented in the Figure 1.19 by changing diameter of circles.

Although the voice call may be terminated, the PS connection remains inactive for a longer time, especially if the user has forgotten that the web browser application is still open in the background of his/her PC or handheld. Since there is currently no data to be transmitted there is no need to occupy any channels, either dedicated or common ones. For such a scenario CELL_PCH and URA_PCH states have been introduced into 3GPP standards.

In RRC CELL_PCH state (as illustrated in Figure 1.20) there is still an active RRC connection (RRC context data stored in UE and SRNC), but no permanent radio bearer is assigned. The radio bearer is set up again if uplink or downlink IP data needs to be transmitted. However, the GTP tunnels on IuPS and Gn remain active although there are no IP frames to be transmitted as long as UE is in CELL_PCH state. The reason for this difference is that network elements have billions of tunnel endpoint identifier (TEID) values

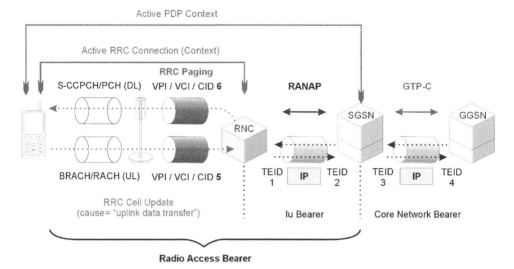

Figure 1.20   Active PDP context – UE in CELL_PCH state

available ($2^{32} = 4\ 294\ 967\ 296$ different values), but – as already pointed out – channelisation codes for dedicated channels on radio interface are always short.

If the UE want to send uplink IP data again it sends an RRC Cell Update Request (cause = 'uplink data transfer') to the SRNC that triggers a new set up of dedicated transport resources. In the case of a downlink data transfer request the UE is paged by the SRNC and answers with an RRC Cell Update Message (cause = 'paging response').

To prevent high frequency of RRC mobility management Cell Update procedures UEs are ordered to enter URA_PCH state. This is useful for fast traffic mobility scenarios, e.g. if the UE is located in a moving car or train. In URA_PCH state the UE only needs to update its current location by performing the RRC URA Update procedure with the SRNC, and an URA is defined as a cluster with an unlimited number of cells.

Since the CELL_PCH state is not implemented in early releases of UE/RNC software mobile phones are ordered to go back to IDLE state instead of CELL_PCH as shown in Figure 1.21. Also in this case the established PDP context remains active, but the RRC connection is terminated and needs to be set up again before uplink data can be transferred or after paging for DL data transfer.

When the UE is ordered to enter IDLE state the GTP tunnel on IuPS is deleted, but the GTP tunnel on Gn remains active, although there are no IP frames transmitted currently.

Also, if the PS connection is inactive the PDP context may remain active for a fairly long time. A typical configuration is to have the PDP context active in SGSN and UE for 48 hours. When the UE is in CELL_PCH, URA_PCH or IDLE state and the network wants to terminate the PDP context due to timer expiry, the UE will be paged by the SGSN to perform the session management deactivate PDP context procedure. Therefore all contexts in UE, RNC and SGSN are deleted and need to be established again at the next call attempt. This is true for both the PDP context between UE and SGSN and for the RRC connection between UE and SRNC.

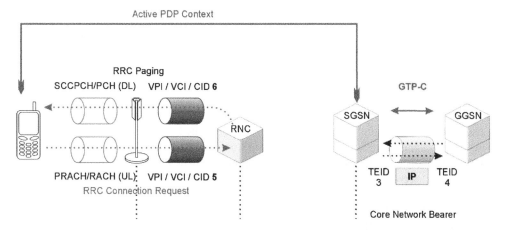

Figure 1.21   Active PDP context – UE in IDLE state

There is another special case of the CELL_DCH state that needs to be discussed briefly. This scenario is possible if the UE as well as the serving cell are HSDPA capable and during a PS connection there is some need to have the highest possible data transmission rates available, e.g. for file download. In such a case for downlink transfer of payload (IP) data the high speed downlink shared channel will be used while the UE is in the CELL_DCH state. In other words: HSDPA call scenarios are characterised by the active PDP context, UE in RRC CELL_DCH state, but the downlink channel mapping is different. Instead of a downlink DCH the HS-DSCH is used. The difference will become evident immediately by looking at Figure 1.22.

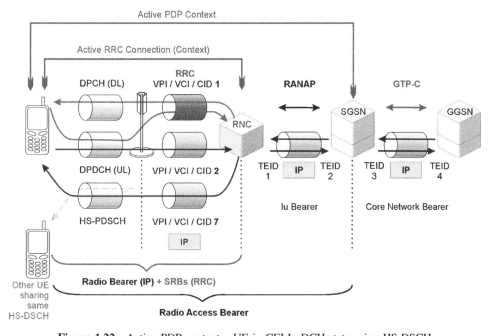

Figure 1.22   Active PDP context – UE in CELL_DCH state using HS-DSCH

The high speed downlink physical shared channel (HS-DPSCH) is used by several UEs simultaneously. That is why it is called a shared channel. Payload belonging to different UEs is identified by a special parameter, the H-RNTI that is unique for each UE in a serving HS-DSCH cell.

For DL payload transport the HS-DSCH is used that is mapped onto the HS-PDSCH. The uplink IP payload is still transferred using a dedicated physical data channel (and appropriate Iub transport bearer) as known from the CELL_DCH scenario. In addition, RRC signalling is exchanged between the UE and RNC using the dedicated channels.

All these channels have to be set up and (re)configured during the call. From a general point of view changes in transport channel mapping to enable/disable HSDPA belong to the group of RRC reconfiguration procedures – just enhanced by another option. And there are also state transitions from CELL_DCH to CELL_FACH and vice versa if the downlink transport channel in CELL_DCH is an HS-DSCH.

A closer look at HSDPA specific call procedures and parameters is given in Chapter 3.

1.2 BASIC ARCHITECTURAL CONCEPT OF PERFORMANCE MEASUREMENT EQUIPMENT BASED ON PROTOCOL ANALYSIS

As already mentioned 3GPP 32.403 describes a performance measurement metrics that can easily be counted, calculated and reported by network elements, but many important factors are missing, especially all kinds of throughput measurement and more complex analysis and correlation of measured events. For this reason a new performance measurement concept (and in turn new equipment) is necessary that fits today's needs of network operators. Since network elements have to focus on switching connections they cannot run additional highly sophisticated performance measurement applications. Hence, network operators have addressed their requirements to the leading manufacturers of protocol analysers and protocol analysis based statistics software. This section will discuss the general basic requirements for such measurement solutions that can be found in all UTRAN performance measurement solutions, no matter which company is the manufacturer.

The first necessary element is a probe that captures data streams from physical links used to transport signalling and payload on Iub, IuCS, IuPS and Iur interfaces.

Once data is captured it needs to be analysed. This can be done directly on the probe or on a separate unit. It is possible to combine data streams captured by multiple probes and send them to a server as shown in Figure 1.23, which executes analysis and stores analysis results in databases. The graphic user interface (GUI) of the performance measurement software then needs to query the database and present query results in tabular or graphic format on the screen. In addition, the GUI may offer different ways to export and import data.

Concepts of how to capture and store data may differ between manufacturers and will not be explained in more detail in this book. Instead a closer look will be given at generic workflow and requirements for data analysis. The following sections describe what is necessary *before* any counter can be counted or any formula can be computed.

As shown in Figure 1.24 once the data stream is captured time stamps are assigned to each frame and then reassembly processes need to be executed. If data streams of multiple probes are combined, all probes need to be highly synchronised. This happens on the ATM level

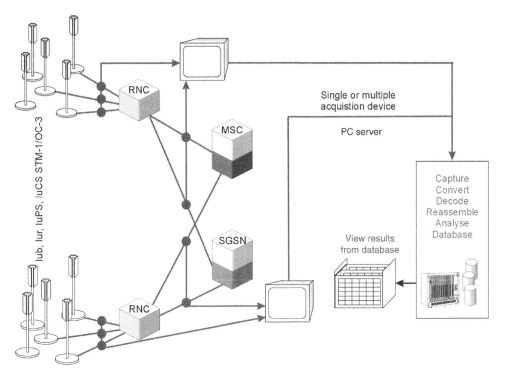

Figure 1.23   Data capture and data analysis

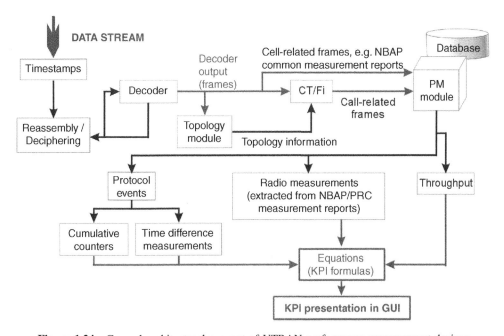

Figure 1.24   General architectural concept of UTRAN performance measurement devices

first, since what was captured on the line are basically ATM cells. Ideally an autoconfiguration application detects if data is transmitted using AAL5 or AAL2 data streams, because different reassembly methods apply for both types. Finally, after the first stage of reassembly, reassembled packets containing bits and bytes are available, which are ready to be decoded. Packets coming out of AAL5 data streams can be decoded using special decoder applications and displayed as SSCOP frames on the monitor screen of the protocol analyser while AAL2 data streams become visible first as either Iub or Iu frame protocol packet data units (PDUs). However, SSCOP and frame protocol decoder output is not enough for a detailed analysis. A look at the protocol stacks shows that for instance Node B Application Part (NBAP), a very interesting signalling protocol, is running on top of SSCOP. Hence, it is necessary to extract the payload field of an SSCOP sequenced data PDU and run another decoding application to see NBAP on the monitor screen.

Frame protocol transport blocks as transmitted via the radio interface can be extracted from Iub. These transport blocks contain radio link control (RLC) data. RLC is the layer 2 protocol on the radio interface that carries RRC/NAS signalling and all kinds of payload (voice, IP). However, to see RRC messages another reassembly process needs to be executed and of course it takes another decoder application to decode the RRC that is encoded using ASN.1 packed encoding rules (PER). ASN.1 PER is a different encoding method compared to methods used for SSCOP and RLC. As a result the architecture of the decoder also needs to be completely different. All in all, this example shows that multiple stages of decoding and reassembly are necessary to get decoder output frames that can be analysed using performance measurement applications. In the case of RLC frames there is another function necessary before decoding takes place, which is deciphering. Deciphering is necessary to make RLC data decodable again, which has been encrypted for transmission via Uu/Iub. Encrypted frames cannot be used for performance measurement purposes (which would otherwise lead to incomplete or incorrect results in higher protocol layer analysis).

It should be noticed that reassembly is not only necessary on the RLC level. Other protocols that have segmentation/reassembly functions are for instance the signalling connection control part (SCCP) used on Iur and IuCS/PS interfaces and the IP used for payload transport across multiple interfaces.

Note: errors in reassembly and deciphering applications are most critical to the following measurement processes. Even if just a single message cannot be reassembled or deciphered correctly this may result in problems of call tracing/filtering and correlation of measurement data.

Correctly decoded frames need to be filtered and sorted. This job is performed by a call trace/filter application (CT/Fi). The topology module runs in parallel to the call trace module, which detects the correlation of ATM virtual connections to UTRAN network elements. In addition the topology application can often display a graphical overview of the network structure and in the future it is possible to combine this topology information with geographical information available in radio network planning tools or location intelligence software like MapInfo software. Topology information will later help to display performance measurement data correlated to network elements and paths.

Not all frames analysed need to be filtered by the call tracer, because not all data is explicitly call-related. A typical example of such data are NBAP common measurement procedures that are reported by each cell of the network to its controlling RNC (CRNC). Because there is a fixed cell-CRNC relation for these procedures, they do not need to be filtered by a call tracer. However, since especially measurement reports need to be filtered per cell it is possible that another special application is introduced at this point that could be described as cell trace. A cell trace application may also extract cell-related information from single calls and could hence be combined with the call trace/filter module. Call trace and cell trace applications also offer excellent opportunities to assign topology information to frames belonging to single calls or single cells. The following performance measurement module (PM module) is the right place to analyse data. Here core functions of counters, gauges and specific generated measurements can be found. Generated measurements are more complex sub-applications to measure e.g. throughput. Measurement results are once again filtered and sorted before they are stored in a database. Filtering and sorting algorithms in PM module are defined by aggregation levels and dimensions required for data presentation later on.

Finally the following basic measurements are available:

- Measurements based on protocol events, which are cumulative counter values and gauges taken from timestamp-oriented time difference measurements.
- Radio-related measurement data extracted from RRC and NBAP measurement reports.
- Throughput measurement data.

This data can be used to calculate different KPIs. KPI calculation is either done in PM module or in a separate application. Calculation results are filtered according to the available topology information and dimensions and then presented on the monitor screen.

That is the way it generally works. However, some of the processes described have such a high impact on finally presented measurement results that it is worthwhile explaining them in more detail in the next sections.

1.2.1 PROTOCOL DECODING AND PROTOCOL STACKS

To decode protocol data correctly it is necessary to have the right protocol stacks. This section provides one example. A complete overview of all UTRAN protocol stacks can be found in Kreher and Ruedebusch (2005).

Figure 1.25 shows protocol stacks on the radio interface (Uu) and Iub. Physical transmission media are ATM cell streams on Iub and wideband CDMA (WCDMA) physical

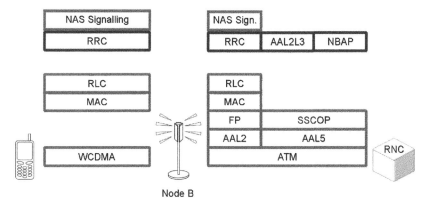

Figure 1.25   Protocol stack Uu/Iub control plane

channels on the Uu interface. Layer 2 functions on the radio interface are embedded in the medium access control (MAC) protocol, which is responsible for channel mapping, e.g. maps a dedicated traffic channel that carries IP data to appropriate transport channels that fit to estimated data transmission rate: very high downlink traffic ⇨ HS-DSCH, high downlink traffic ⇨ DCH, low downlink traffic ⇨ FACH. The RLC is responsible for error detection and error correction (when using acknowledged mode) as well as for security (integrity protection and ciphering). Payload transmitted in RLC transport blocks contains RRC signalling and optionally NAS signalling for mobility management, session management and call control within the control plane. Voice and IP data, often called payload, is transmitted by the RLC/MAC on the user plane. User plane and control plane information is mostly transmitted using different transport channels (only possible exception: RACH/FACH). The frame protocol (FP) present on Iub only is necessary to forward RLC transport blocks built by UE and RNC to/from Node B. The need for this special frame protocol is due to the fact that the RLC has different transport formats using different transport block sizes and time transmission intervals (TTI). Since Node B and its cells are quite simple hardware that is not able to make any frame processing, but just fulfils the function of an interface converter and multiplexer (as described in the section about PS call analysis), frames need to be forwarded on Iub transparently in the same transport format as required for radio interface transmission. This is a main job of the frame protocol and (as explained later) there is also some radio-related measurement data embedded in uplink FP frames.

The function of different higher layer signalling protocols can be best explained looking at Figure 1.26, which shows channel types and related protocols.

Here it becomes clear that the radio resource control (RRC) is, so to speak, the way of expression between UE and RNC while NBAP is the way of expression between Node B and RNC. NAS protocols are ways of expression between UE and core network elements MSC and SGSN. UMTS logical channels carry higher layer signalling (RRC/NAS) or voice/data while on transport channels only RLC transport blocks can be found. Thanks to services provided by FP, transport channels are end-to-end connections between UE and RNC

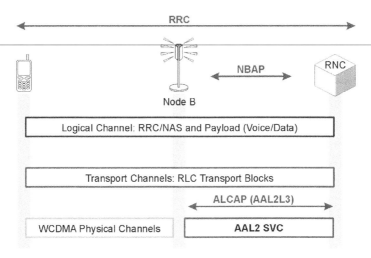

Figure 1.26 UMTS channels and protocol functions

although the physical transport medium is changed right in the middle from AAL2 switched virtual connection (SVC) on Iub to physical channels sent/received by cell and UE antennas.

If dedicated physical channels are used for communication on the radio interface, the necessary resources (= codes) need to be assigned by the RNC. On Iub AAL2 SVCs that carry dedicated transport channels need to be set up. Set up and release of AAL2 SVCs is the task of the ALCAP function. In UTRAN configurations following 3GPP standards, the ALCAP function is provided by a protocol layer called AAL2L3 (ATM adaptation layer type 2 – layer 3 signalling). However, there are also proprietary NBAP versions on the market that have an embedded ALCAP function, which means that NBAP is able to set up and delete AAL2 SVCs on its own without the help of the AAL2L3 protocol layer.

A different protocol version needs different decoders and also special call trace/filter modifications and different event counter definitions because counter definition criteria are based on decoded protocol information. This is the first important point one needs to know about decoding when discussing performance measurement issues.

The second important point regarding decoding is that UTRAN protocol standards are frequently updated. There are four different releases of UMTS standards by 3GPP: Release 99 (also called Rel. 3), Release 4, 5 and 6. The standardisation process regarding release features has been completed for Rel. 3 and 4, but is still ongoing for Rel. 5 and 6. However, even if the feature definition is complete, protocol definitions may change. Usually quarterly there is a new version of important UMTS protocols. The hierarchy of protocol versions is as shown in the Figure 1.27.

In total, four different protocol version standard documents in each of the four different releases are published each year, which means there are 16 different versions of NBAP, 16 different versions of RRC, RANAP, RNSAP etc. Each version requires its own decoder application otherwise there is a risk of decoding errors. Even if the protocol running on a defined network path can be detected by autoconfiguration applications, the appropriate protocol version requires an appropriate decoder module and configuration input by the operator, who needs to configure which protocol version is used to ensure correct decoding.

1.2.2 DIVERSITY COMBINING AND FILTERING

It has already been explained how important reassembly and filter procedures are to ensure precise performance measurement. The special conditions of UTRAN FDD mode require another look at them to investigate some special cases of data stream combining that are necessary if identical data is transmitted via several radio and UTRAN interfaces.

All previous statements regarding RLC reassembly are valid as long as the UE is in CELL_DCH mode and only one radio link belongs to the active set as shown in the Figure 1.28.

This scenario is characterised by the fact that bidirectional transfer of RLC transport blocks on a single Uu/Iub interface connection is enabled. Radio connection is served by a single cell and each transport block sent via this single cell can be directly monitored without a special combining/filtering mechanism on Iub. If no softer or soft handover or transport channel type switching is performed this situation remains unchanged until the end of the call.

Figure 1.28 shows an example for the transport of an RRC measurement report message sent by UE to SRNC. The contents determine the length of the message. It depends on the

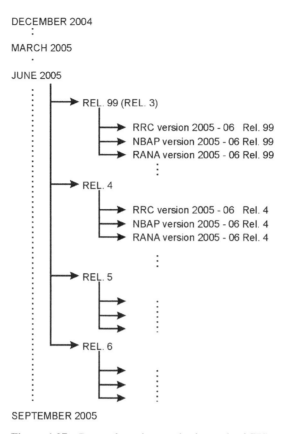

Figure 1.27   Protocol versions and releases by 3GPP

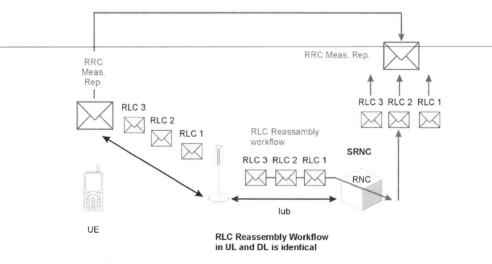

Figure 1.28   RLC reassembly – UE in CELL_DCH, one radio link in active set

total length of the RRC message how many RLC transport blocks are necessary to transport the whole RRC message. In the example it is assumed that the number of transport blocks necessary for transmission of this message is three. These three RLC transport blocks – each of which has its unique sequence number – are sent in sequence, first on the radio interface and then on the Iub physical transport bearer (AAL2 SVC). Only the RLC transport blocks actually exist on both interfaces and the original RRC measurement report message is not reassembled before transport blocks have been received by SRNC. However, the protocol analysers that capture these transport blocks on Iub – no matter if these analysers are equipped with performance measurement applications or not – need to reassemble the original RRC message in the same way as it is done by SRNC. By the way, this is true for both uplink and downlink RRC messages.

Now assume that the call goes into a soft handover situation, in which a second cell belonging to a different Node B provides a second radio link that is part of the active set for reasons explained in Figure 1.7.

Imagine that the same RRC measurement report message is sent by the UE again. Now – as shown in Figure 1.29 – the same three RLC transport blocks containing segments of this message are sent twice. Each transport block is sent simultaneously on two different radio links. Since two different Node Bs are involved in this scenario these identical transport blocks can also be monitored on two different Iub interfaces – except there is one – and only one! – RRC measurement report that is sent in active connection between UE and SRNC.

Two different radio links (= two different cells) have been used for the transport of identical RLC transport blocks. This means, in the case of different radio transmission conditions, it must be expected that on individual physical channels of each cell individual bit errors will occur. The uplink bit error rate is a radio interface KPI that is mostly estimated from the error rate of pilot bits sent on the uplink dedicated physical control channel. Node B then encodes this measured bit error rate as a quality estimate value, which is appended to

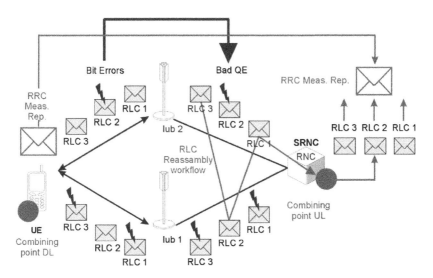

Figure 1.29   UE in CELL_DCH – two radio links in active set – soft handover

every transport block set sent by the FP in the uplink direction. If two identical RLC blocks are received by SRNC the radio network controller decides from the quality estimate which of the two identical blocks is assumed to have fewer bit errors and is hence used for reassembly. The other block is discarded. If a transport block has been protected by a cyclic redundancy check and a CRC error has been indicated for this block, SRNC usually discards this block anyway, no matter which bit error rate has been measured. Instead of the quality estimate other uplink radio quality parameters can be used to determine which of the two or more identical transport blocks is expected to be the best one in the uplink direction. The signal-to-interference ratio (SIR) reported by NBAP for each uplink dedicated physical channel is a good alternative decision maker, but there is no explicit definition by 3GPP of how the quality of uplink frames is acquired.

This procedure itself is officially called selection combining (3GPP 25.401), but it is also known as macro-diversity combining. It applies to all uplink data transmissions in the case of soft handover situations. In downlink there is also a combination of data streams on the UE side, but this maximum ratio combining follows different rules and is explained in Figure 1.30. Macro-diversity is a feature that is nowadays only available in SRNC as far as the author knows (3GPP offers the possibility to implement selection combining in DRNC also) and it only applies to uplink data streams, no matter if they carry RRC signalling or payload (voice/data).

How radio link combining is realised on the downlink is seen when analysing a special case of soft handover that is called softer handover. In this case cells that provide radio links belonging to the same active set are controlled by the same Node B as shown in Figure 1.30.

In softer handover the uplink and downlink radio signals are combined behind the rake receiver antenna of UE or Node B using maximum ratio combining. The rake receiver is a special antenna design that allows the device to send/receive the same data simultaneously on several radio channels. It is now possible to add received signal levels coming from different radio links, when it is known that the received data is identical. The higher the

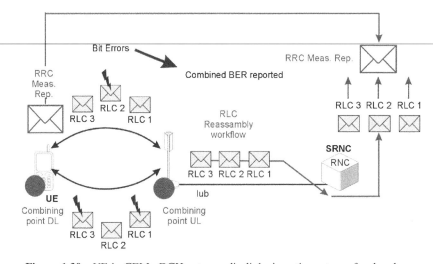

Figure 1.30   UE in CELL_DCH – two radio links in active set – softer handover

resulting signal level the better signal-to-noise ratio can be achieved. Based on this effect, it might be possible that two different radio links, which would be not good enough for signal transmission in stand-alone mode, become good enough to be used in softer handover due to this so-called micro-diversity combining (a synonym for maximum ratio combining of uplink data streams in Node B).

However, since signals are already merged right behind receiving antennas there is just one uplink/data stream monitored on Iub while there are two independent data streams on the radio interface. The quality estimate attached to uplink transport blocks also represents a combined value, an average of bit error rate measured on two (or more) radio links, and as a result it is not possible to find out on which radio link of the active set more bit errors occurred. This means that in this scenario it is no longer possible to measure on Iub which of the radio links of active connection has been better or worse. This limitation is quite important, because it disables the correlation of some performance-related data monitored on Iub to a single cell (example: uplink bit error rate).

In the future it is planned to introduce another feature that has an impact on RLC reassembly processes using downlink data transmission. This feature is shown in Figure 1.31 and is called site selection diversity transmission (SSDT) and it reduces the overhead of downlink data transfer due to softer and soft handover scenarios. Basically it is defined that with SSDT only the best of all radio links in the active set will be used for transmission of downlink data.

While in soft handover there are still identical uplink RLC frames (containing the same RRC message) sent on the different Iub interfaces in uplink direction there is only *one* Iub interface/*one* radio link used for downlink data transmission. The single downlink radio link is the one with the best quality. The quality indicator in this case is the feedback information (FBI) transmitted on the uplink dedicated physical control channel (DPCCH).

An indicator in the NBAP Radio Link Setup Request message shows if SSDT can be enabled for a defined radio link or not. While the feature is currently not used yet

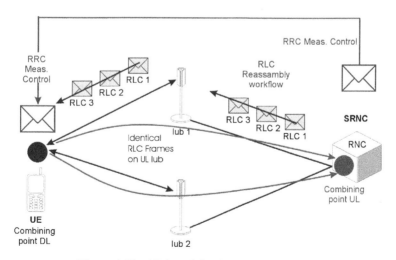

Figure 1.31   UE in soft handover, SSDT enabled

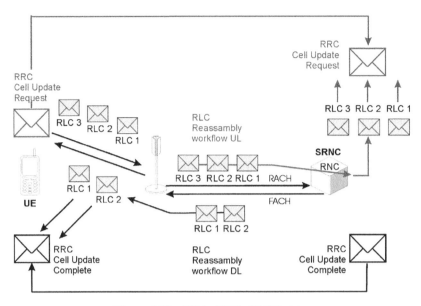

Figure 1.32   UE in CELL_FACH state

(May 2006) it is expected to become important for Iub transport resource optimisation in later deployment stages of UMTS networks.

This section ends by having a final look at the CELL_FACH state illustrated in Figure 1.32. Here neither macro-diversity nor micro-diversity combining applies, but uplink and downlink transport blocks are sent on different Iub physical transport bearers (different VPI/VCI/CID).

This is not a problem for protocol analysers except for the fact that the used transport channel can be changed if the UE is ordered to go back to CELL_DCH state right in the middle of transmitting RLC transport blocks belonging to the same higher layer message. Therefore it is possible that the first RLC transport block of an RRC message transmitted in the uplink direction is sent on RACH while the second transport block that carries a fragment of the same RRC message is sent on DCH (and hence, on different Iub VPI/VCI/ CID again). Switching of the transport channel is indicated if the RRC Cell Update Complete message shown in the lower part of Figure 1.32 contains the RRC state indicator = 'CELL_DCH'.

1.2.3 STATE TRANSITION ANALYSIS

A very precise method to analyse how a call develops step by step is to run a call state analysis using a state machine. It has already been mentioned that RRC entities in UE and SRNC run a state machine for each connection. However, it is only a minor subset of a greater framework to trace state transitions in these network element state machines. Using a proprietary definition it is possible to define states that describe each single procedure of a call including optional ones. In such models signalling events trigger state transitions and, if a call fails, the root cause can be found very quickly.

It is not a new invention to split up a call in single procedures and define states based on call progress. ITU-T introduces such call progress analysis in basic call state models (BCSM) of Intelligent Network Application Part (INAP) standard definitions. Having a look at BCSM of INAP Capability Set 1 (CS-1) and CS-2 it also becomes clear that the more triggers and states (in INAP called *points in call*) are defined the more detailed and better the analysis is.

Using state transition models it is not only possible to analyse the progress of a call, but also to follow up resource allocation. Remember the analysis of PS calls. Based on the RRC state it can be determined if common transport channels RACH and FACH (RRC state 'CELL_FACH') or dedicated transport channels DCH (RRC state 'CELL_DCH') are used in a certain phase of a PS connection. Trigger events for transitions between CELL_DCH and CELL_FACH are found directly in RRC signalling messages and the whole RRC state machine including all definitions of state transitions is described in detail in 3GPP 25.331. However, remember that the maximum possible throughput in CELL_DCH depends on the assigned spreading factor of the channelisation code. Using proprietary definitions it is now possible to define sub-states of CELL_DCH according to different spreading factors.

Figure 1.33 shows the set up of a PDP context and its appropriate initial radio bearer using the channelisation code # 3/spreading factor 32 (code # in line of message, SF in column header). Note, the start up of an RRC connection and the PDP context could also be analysed using a set of states, e.g. if Attach Request (ATRQ) is received the call state is transitioned to *Attach_Attempt* and if Attach Accept is monitored the next call state *Attach_Success* is entered and the state machine waits for an Activate PDP Context Request (APCR) to enter the next state 'PDP Context Activation Attempted'. If Attach Reject is monitored the final state of this call would be *Attach_Failure* and the record can be written to a database that contains all final states of all calls for further analysis.

In this example a radio bearer is set up and the UE is ordered to jump into RRC state CELL_DCH using spreading factor 32. However, after a few seconds the network already starts a reconfiguration procedure and sends the UE into CELL_FACH state (no spreading factor assigned in this case).

After 46 seconds in CELL_FACH another reconfiguration is triggered by the RRC Measurement Report containing event 4A (higher data volume to be transported in uplink

Long Time	3. Prot	3. MSG	Procedure Code	4. Prot	4. MSG	rrc-stateindicator	sf32	sf128
10:06:13,363,263	RRC_CCCH_UL	rrcConnectionRequest						
10:06:13,454,737	RRC_CCCH_DL	rrcConnectionSetup				cell-FACH		
10:06:14,236,420	RRC_DCCH_UL	rrcConnectionSetupComplete						
10:06:14,808,574	RRC_DCCH_DL	initialDirectTransfer		GMM-DMTAP	ATRQ			
10:06:14,935,146	RRC_DCCH_DL	securityModeCommand						
10:06:15,644,397	RRC_DCCH_UL	securityModeComplete						
10:06:15,744,913	RRC_DCCH_DL	downlinkDirectTransfer		GMM-DMTAP	ATAC			
10:06:16,245,719	RRC_DCCH_UL	uplinkDirectTransfer		GMM-DMTAP	ACOM			
10:06:17,444,376	RRC_DCCH_UL	uplinkDirectTransfer		GSM-DMTAP	APCR			
10:06:17,665,740	**NBAP**	**initiatingMessage**	**id-radioLinkSetup**					
10:06:17,776,781	**NBAP**	**succesfulOutcome**	**id-radioLinkSetup**					
10:06:17,789,893	**AAL2L3**	**ERQ**						
10:06:17,809,253	**AAL2L3**	**ECF**						
10:06:17,925,033	**AAL2L3**	**ERQ**						
10:06:17,944,781	**AAL2L3**	**ECF**						
10:06:18,084,867	RRC_DCCH_DL	radioBearerSetup				cell-DCH	3	
10:06:19,030,468	RRC_DCCH_UL	radioBearerSetupComplete						
10:06:19,256,284	RRC_DCCH_DL	measurementControl						
10:06:19,416,233	RRC_DCCH_DL	downlinkDirectTransfer		GSM-DMTAP	APCA			
10:06:19,576,164	RRC_DCCH_DL	measurementControl						
10:06:19,896,124	RRC_DCCH_DL	measurementControl						
10:06:20,616,252	RRC_DCCH_DL	measurementControl						
10:06:21,136,136	RRC_DCCH_DL	physicalChannelReconfiguration				cell-FACH		
10:06:23,000,391	RRC_CCCH_UL	cellUpdate						
10:06:23,144,852	RRC_DCCH_DL	cellUpdateConfirm				cell-FACH		

Figure 1.33 PS call state analysis – Part 1

Long Time	3. Prot	3. MSG	Procedure Code	4. Prot	4. MSG	trafficVo...	rrc-statei...	sF32 sF128
10:06:21,136,136	RRC_DCCH_DL	physicalChannelReconfiguration					cell-FACH	
10:06:23,800,391	RRC_CCCH_UL	cellUpdate						
10:06:23,144,852	RRC_DCCH_DL	cellUpdateConfirm					cell-FACH	
10:06:23,545,810	RRC_DCCH_UL	physicalChannelReconfigurationComplete						
10:06:23,545,810	RRC_DCCH_UL	utranMobilityInformationConfirm						
10:06:23,614,729	MBAP	initiatingMessage	id-radioLinkDeletion					
10:06:23,685,324	MBAP	succesfulOutcome	id-radioLinkDeletion					
10:06:23,694,937	AAL2L3	REL						
10:06:23,695,132	AAL2L3	REL						
10:06:23,713,616	AAL2L3	RLC						
10:06:23,720,227	AAL2L3	RLC						
10:07:06,763,866	RRC_DCCH_UL	measurementReport				e4a		
10:07:06,870,923	MBAP	initiatingMessage	id-radioLinkSetup					
10:07:06,936,825	MBAP	succesfulOutcome	id-radioLinkSetup					
10:07:06,955,188	AAL2L3	ERQ						
10:07:06,955,576	AAL2L3	ERQ						
10:07:06,982,519	AAL2L3	ECF						
10:07:06,989,131	AAL2L3	ECF						
10:07:07,174,424	RRC_DCCH_DL	transportChannelReconfiguration					cell-DCH	3
10:07:07,951,051	RRC_DCCH_UL	transportChannelReconfigurationComplete						
10:07:08,135,474	RRC_DCCH_DL	measurementControl						
10:07:08,455,527	RRC_DCCH_DL	measurementControl						
10:07:08,749,345	RRC_DCCH_UL	measurementReport				e4b		
10:07:08,775,486	RRC_DCCH_DL	measurementControl						
10:07:10,615,424	RRC_DCCH_DL	transportChannelReconfiguration					cell-DCH	2
10:07:11,269,311	RRC_DCCH_UL	transportChannelReconfigurationComplete						
10:07:11,415,521	RRC_DCCH_DL	measurementControl						

Figure 1.34 PS call state analysis – Part 2

direction) as shown in Figure 1.34. Once again dedicated transport channels are set up and the UE is ordered to go back to CELL_DCH using code with spreading factor 32.

Due to low uplink data volume indicated by event 4B UE channelisation code #3/SF 32 is substituted by code #2/SF 128 after just 3 seconds.

This change between code # 2/SF 128 and code #3 /SF 32 is occurs several times in fairly short time intervals (see Figure 1.35). The maximum theoretical throughput levels assigned during the call can then be displayed in a time line diagram as show in Figure 1.36.

The analysed thresholds (based on defined states) can now be used as input for further call analysis, e.g. to compute how often and how fast assigned resources are reconfigured or to compare measured throughput with a theoretical possible maximum in different states of call. However, the main limitation of a state machine is that it can only be used to analyse signalling procedures. An in-depth user plane analysis, e.g. measuring throughput or UL BLER (see Section 2.1) cannot be realised using a state machine and requires another separate application in the performance measurement software.

Long Time	3. Prot	3. MSG	Procedure Code	4. Prot	4. MSG	trafficVo...	rrc-statei...	sF32 sF128
10:07:12,975,555	RRC_DCCH_DL	transportChannelReconfiguration					cell-DCH	3
10:07:13,550,056	RRC_DCCH_UL	transportChannelReconfigurationComplete						
10:07:13,695,391	RRC_DCCH_DL	measurementControl						
10:07:15,229,965	RRC_DCCH_UL	measurementReport				e4b		
10:07:16,215,596	RRC_DCCH_DL	transportChannelReconfiguration					cell-DCH	2
10:07:16,829,860	RRC_DCCH_UL	transportChannelReconfigurationComplete						
10:07:16,975,390	RRC_DCCH_DL	measurementControl						
10:07:18,535,424	RRC_DCCH_DL	transportChannelReconfiguration					cell-DCH	3
10:07:19,109,925	RRC_DCCH_UL	transportChannelReconfigurationComplete						
10:07:19,255,552	RRC_DCCH_DL	measurementControl						
10:07:20,789,876	RRC_DCCH_UL	measurementReport				e4b		
10:07:21,775,409	RRC_DCCH_DL	transportChannelReconfiguration					cell-DCH	2
10:07:22,389,770	RRC_DCCH_UL	transportChannelReconfigurationComplete						
10:07:22,535,593	RRC_DCCH_DL	measurementControl						
10:07:24,095,529	RRC_DCCH_DL	transportChannelReconfiguration					cell-DCH	3
10:07:24,669,835	RRC_DCCH_UL	transportChannelReconfigurationComplete						
10:07:24,815,365	RRC_DCCH_DL	measurementControl						
10:07:26,351,808	RRC_DCCH_UL	measurementReport				e4b		
10:07:27,335,278	RRC_DCCH_DL	transportChannelReconfiguration					cell-DCH	2
10:07:27,989,598	RRC_DCCH_UL	transportChannelReconfigurationComplete						
10:07:28,135,419	RRC_DCCH_DL	measurementControl						
10:07:29,695,356	RRC_DCCH_DL	transportChannelReconfiguration					cell-DCH	3
10:07:30,269,856	RRC_DCCH_UL	transportChannelReconfigurationComplete						
10:07:30,415,484	RRC_DCCH_DL	measurementControl						
10:07:31,949,817	RRC_DCCH_UL	measurementReport				e4b		
10:07:33,935,382	RRC_DCCH_DL	transportChannelReconfiguration					cell-DCH	2
10:07:34,589,604	RRC_DCCH_UL	transportChannelReconfigurationComplete						

Figure 1.35 PS call state analysis – Part 3

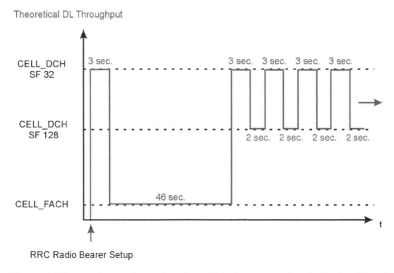

Figure 1.36   Maximum theoretical downlink throughput thresholds for PS call

1.3 AGGREGATION LEVELS/DIMENSIONS

It is now time to talk about correlation of measurement data to network element and topology parameters again – this time in detail.

A dimension is a special filtering of performance measurement data to build subsets of events or show collected data on different aggregation levels such as cell, RNC, SGSN etc.

Some dimensions/aggregation levels are given by the nature of measurement procedures and information elements in signalling messages by following 3GPP, others by following abstract proprietary definitions and correlation processes. This section gives an overview of the most common aggregation levels, possible limitations of accuracy and useful combinations with other dimensions.

Note: the following overview is far from complete. There are many more customised aggregation levels defined by operators and often bound to special needs of network planning departments and operation/maintenance groups that cannot even be mentioned in this book. To put it in a nutshell, there is no limitation on defining additional dimensions.

1.3.1 SGSN DIMENSION

All event counters that deal with the analysis of GPRS mobility management and session management can be easily aggregated on SGSN level without any limitation. The same is true for user data stream analysis derived from Iu bearers (GTP-U tunnels on IuPS). Typical gauges like minimum/maximum/mean RAB Setup Time can be correlated to SGSN, but actually it is up to the SRNC to set up RABs. A combination with IMSI (that can always be detected on Iu from the RANAP Common ID message) is often useful, because it allows the identification of procedures correlated to roaming UEs. In addition, all information elements found in NAS signalling and RANAP messages can be used for dimension filtering, especially access point name (APN) for analysis of PDP context activation, traffic class and service area identity (SAI). SAI is often in a 1 : 1 relationship with the cell from where the call started. Hence, it allows a correlation of RANAP/NAS events to cell level. However, due to RRC mobility, the UE changes cell very quickly and if e.g. the RAB drops, it is very unlikely that the UE is still located in the cell identified by the SAI. Furthermore, filtering per SRNC could provide good analysis results as well.

1.3.2 MSC DIMENSION

All event counters that deal with analysis of CS mobility management and call control can be aggregated on MSC level without limitations. Additionally, ALCAP events can be correlated to MSC to analyse set up, modification and release of physical IuCS user plane transport bearers (AAL2 SVC). All statements for SGSN aggregation level regarding additional filtering using traffic class, IMSI and SAI apply to MSC in the same way.

1.3.3 SRNC DIMENSION

The serving RNC aggregation level is meaningful for all analysis of protocol events derived from RRC, RANAP, RNSAP and ALCAP signalling. User plane analysis for all radio bearers can also be shown on SRNC level. NAS events and measurements can be shown related to the SRNC, but do not necessarily need to be shown since NAS signalling is transparently forwarded by the SRNC to/from core network domains. For all kinds of inter-RNC handovers/relocations it is useful to see the source RNC and target RNC of the handover procedure or in the case of inter-RAT handover source/target BSC instead of RNC. A perfect tool to realise this requirement is the handover matrix described in Section 2.14.

All RRC signalling messages sent/received on common transport channels RACH and FACH can additionally be correlated to a single cell without limitations if the topology module provides information about which common transport channel belongs to which cell.

As long as a call is not in inter-RNC soft handover the rule applies that SRNC = CRNC and no DRNC is involved.

1.3.4 DRNC DIMENSION

Drift RNC level only applies for analysis of calls performing inter-RNC soft handover. It is recommended to use DRNC aggregation level only for RNSAP signalling events. RRC signalling and user plane traffic monitored on Iur should be shown on SRNC or cell level.

1.3.5 CRNC DIMENSION

Controlling RNC level applies to the analysis of all NBAP measurement reports and protocol events, because NBAP is the way of expression between CRNC and Node B.

It especially makes sense to aggregate all counters for NBAP Class 1 and ALCAP procedures to CRNC. Also all kinds of statistics regarding radio resource allocation should also be correlated to CRNC, because it is CRNC that assigns radio resources.

For NBAP common measurements CRNC can only be seen as assisting dimension, because these measurements come from single cells and therefore would be better displayed at cell level. Due to this, in case of NBAP common measurement reports, correlation with cell identity is mandatory. If cell ID is not available from the monitored signalling correlation to NBAP, the measurement ID is an option as explained more detailed in Section 2.2.2.

Never aggregate radio-related performance measurements extracted from NBAP Common Measurement Report messages on different aggregation levels than Node B and cell level!

1.3.6 NODE B DIMENSION

Usually Node B dimension is not very meaningful on its own, but can become important to assist correlation of cell-related or channel-related measurements and protocol events, especially if cell identity or channel identity has not yet been detected by the topology module. If for instance correlation cell-ID/measurement-ID in NBAP common measurement procedures is not available it makes sense to show reported values on a Node B/measurement-ID aggregation level as described in Section 2.2.2. This allows the user to detect abnormal behaviour reported from the single cell of a Node B – although the real cell-ID still needs to be checked in CRNC settings available in the RNC operation and maintenance centre (OMC).

The only NBAP procedures that require mandatory Node B aggregation level are the NBAP audit and NBAP reset procedures, because they are related to Node B itself and not to single cells or transport channels. For NBAP dedicated measurements, the Node B aggregation level depends on the measurement object type. If the measurement object is defined as 'all radio links within a single Node B communication context' there is no other choice than to show measurement results for Node B when the call is in softer handover situation. The same is true for the bit error rate (quality estimate) due to micro-diversity combining of uplink radio channels in softer handover.

In addition, Node B aggregation level is meaningful for analysis of all procedures dealing with set up, modification and deletion of Iub physical transport bearers (AAL2 SVCs) using ALCAP signalling, because the ALCAP entity is bound to Node B, not to cells.

1.3.7 CELL DIMENSION

All RRC messages sent on common transport channels like RACH and FACH can be correlated to a single cell without limitation. The same is true for PS payload streams sent on HS-DSCH. However, it would be beneficial to define a special cell type for HS-DSCH analysis, which should be named according to 3GPP as the *serving HS-DSCH cell.*

All RRC messages sent on DCH are difficult to correlate to single cells because in FDD mode the call tends to be in softer or soft handover situation. In average the call is in soft handover in 50% of total call duration and one out of four (25%) soft handover situations belong to the special group of softer handover. It is important to remember that UE has only one active RRC connection with SRNC – no matter how many radio links (= cells) belong to the active set. The same situation applies in user plane data streams.

Aggregating RRC counters to all cells belonging to an active set leads to an incorrect analysis.

To overcome this problem some NEM and network operators have defined a best cell aggregation level. The idea is to show RRC events as well as selected data stream analysis as user perceived throughput related to the best cell of connection. As long as there is just one cell in the active set or if UE is in CELL_FACH state a single serving cell is the best cell of connection. Then, after radio link addition the best cell can be analysed based on CPICH Ec/N0 values reported for all cells of the active set. This procedure requires a special application that tracks changes of the active set of each UE and frequently updates changes of the best cell based on RRC measurement reports sent by the UE. And as another limitation it must be kept in mind that the cell with the best CPICH Ec/N0 reported value is the best cell for downlink data transmission. However, on uplink paths the situation might be different due to the fact that in FDD mode uplink and downlink channels are transmitted using different frequency ranges. In a typical interference scenario sidebands of other radio devices can influence UTRAN frequency bands. They only interfere with a defined frequency range and may hit only the uplink, for instance, but not the downlink band of a UTRAN cell as shown in Figure 1.37.

In such a case the cell that reports the best CPICH Ec/N0 would indeed be the best cell for downlink radio transmission, while simultaneously it could be the worst cell for uplink radio transmission. These effects become visible if extreme differences in uplink and downlink radio quality parameters such as uplink and downlink block error rate (BLER) are displayed for the same cell.

In addition, it must be kept in mind that the best cell cannot be determined only by Ec/N0, but by other radio quality measurement results, too. Pathloss measurement, for instance is a possible alternative parameter if it is reported by the UE to SRNC and while the best cell is the cell having the highest reported Ec/N0 value and is by definition placed on top of the RRC intra-frequency measured result list, it is the cell with the lowest reported pathloss values and by definition placed at the end of the intra-frequency measured result list. It is also possible to determine the best cell based on uplink quality measurement results such as the uplink signal-interference ratio (SIR) reported using the NBAP dedicated measurement report. Having all these options in mind there is no doubt that the best cell determination is based on proprietary definitions and also depends on available reported measurement values.

For hard handovers, the source cell and target cell are of interest for analysis. Usually the source cell is the cell in which the hard handover command (= Attempt) has been monitored,

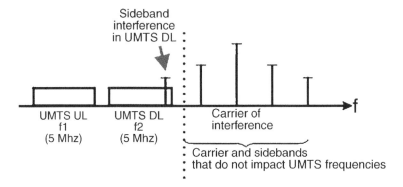

Figure 1.37   Interference caused by sidebands

while the success event (hard handover complete) is monitored in the target cell. To compute a handover ratio both events need to be counted for the same cell, which means either the source or target cell.

It is meaningful to measure RLC throughput on single cell level if more than one cell is in the active set, because RLC throughput gives an impression of how much load is on each cell. In WCDMA the capacity of a cell depends directly on cell load.

Softer and soft handover require special rules for cell level analysis since there is no source and target cell, but radio link additions and radio link deletions. An analysis concerning which cells are often seen together in active sets can be done using a cell neighbour matrix with special filter conditions (see Section 2.14).

As long as dedicated NBAP measurement procedures report the quality of a single radio link they also can be correlated to a single cell without limitations. However, if the same measurement procedures apply to a radio link set or multiple radio links within the same Node B a correlation on cell level cannot be done.

If NBAP dedicated measurement procedures apply for single radio links, radio link sets or all radio links within a single Node B, the communication context can be detected from the dedicated measurement object type sequence in the NBAP dedicated measurement initialisation message. If performance measurement software is not able to distinguish clearly between these options it is better to show measurement values derived from NBAP dedicated measurement reports on call level instead of cell level. Otherwise there may be incorrect statements regarding the quality of single radio links/cells.

1.3.8 CALL/CONNECTION DIMENSION

For call (for all statements in this section it is assumed call = connection) set up, modification and call drop/call release, call aggregation level is mandatory. Involved SRNC, DRNC, CRNC and core network elements may be shown as well. A correlation between call and cell is often difficult due to reasons described in the previous section.

For soft and softer handover analysis it is important to track all changes of the active set of a single call so that one knows which and how many radio links are involved at any time (in call). For correlation with user plane analysis (e.g. throughput), settings and changes of transport format sets for each radio bearer must be known. Hence, radio bearer aggregation level could be a good subset filter for user plane analysis.

There can also be good correlation between call and traffic class as well as between call and type of service (call type).

1.3.9 UE DIMENSIONS

No matter if permanent or temporary UE identifiers are used – they are always good for analysis of NAS signalling, RRC events and user plane. Assigned PDP addresses for PS services could be handled as UE identifiers, too, although they are often not directly assigned to UE, but to an application running on the mobile phone or PC connected to the mobile. A correlation of NBAP dedicated events and measurement values with UE IDs may also make sense, but needs to be discussed in each single case.

It is also often useful to check not only IMSI or TMSI, but also International Mobile Equipment Identity (IMEI), because in live networks it has been observed that for certain UE manufacturers/types handover procedures are triggered faster if the handover decision is made by the RNC based on uplink quality measurements. Due to this it is possible to see a much higher UMTS-to-GSM inter-RAT handover/relocation rate for some groups of IMEI than for others, however, the CS call drop rate in UTRAN for these IMEIs is lower. The UE type is encoded in that part of the IMEI that is known as type allocation code (TAC). The TAC consists of eight bits. The leading six bits are enough to identify a certain type of UE, e.g. a Nokia 6680 mobile phone.

1.3.10 RADIO BEARER/RADIO ACCESS BEARER TYPE DIMENSIONS

The maximum bit rate that is possible for a certain type of radio bearer or radio access bearer is mandatory for transport channel and service throughput analysis as a reference value. Otherwise no statement is possible regarding the quality of throughput measurement results compared to network capacities.

The radio access bearer is a service for the *transport of user data (voice, IP payload etc.) between UE and core network domains (MSC or MGW, SGSN) provided by the serving RNC.*

RAB parameters are monitored in RANAP signalling and provide a more abstract view of quality of service provided for a single connection than radio bearer parameters monitored in RRC signalling. Radio bearer and Iu bearer represent two halves of a radio access bearer.

The radio bearer is a service for transport of user data (voice, IP payload etc.) between UE and the serving RNC.

RAB maximum bit rate parameters can be found in RANAP RAB Assignment Request message as shown in Figure 1.38 (mandatory parameter) and in RANAP RAB Assignment Response (optional parameter). The difference is that the request message contains only the

```
TS 25.413 V5.9.0 (RANAP)  initiatingMessage (= initiatingMessage)
ranapPDU
1 initiatingMessage
1.1 procedureCode                              id-RAB-Assignment
1.2 criticality                                reject
1.3 value
1.3.1 protocolIEs
1.3.1.1 sequence
1.3.1.1.1 id                                   id-RAB-SetupOrModifyList
1.3.1.1.2 criticality                          ignore
1.3.1.1.3 value
1.3.1.1.3.1 sequenceOf
1.3.1.1.3.1.1 sequence
1.3.1.1.3.1.1.1 id                             id-RAB-SetupOrModifyItem
1.3.1.1.3.1.1.2 firstCriticality               reject
1.3.1.1.3.1.1.3 firstValue
1.3.1.1.3.1.1.3.1 rAB-ID                       05
1.3.1.1.3.1.1.3.2 rAB-Parameters
1.3.1.1.3.1.1.3.2.1 trafficClass               background
1.3.1.1.3.1.1.3.2.2 rAB-AsymmetryIndicator     asymmetric-bidirectional
1.3.1.1.3.1.1.3.2.3 maxBitrate
1.3.1.1.3.1.1.3.2.3.1 maxBitrate               128000
1.3.1.1.3.1.1.3.2.3.2 maxBitrate               64000
1.3.1.1.3.1.1.3.2.4 deliveryOrder              delivery-order-not-requested
1.3.1.1.3.1.1.3.2.5 maxSDU-Size                12000
1.3.1.1.3.1.1.3.2.6 sDU-Parameters
```

Figure 1.38 Maximum bit rate (DL/UL: 128/64 kbps) in RANAP RAB Assignment Request

Long Time	2. Prot	2. MSG	3. Prot	3. MSG	Procedure Code
11:33:28,004,959	RLC/MAC	AM DATA DCH	RRC_DCCH_UL	uplinkDirectTransfer	
11:33:28,112,232	SSCOP	SD	RL	RL	id-RAB-Assignment
11:33:28,121,041	SSCOP	SD	NBAP	initiatingMessage	id-synchronisedRadioLinkReconfigurationPreparation
11:33:28,493,530	RLC/MAC	AM DATA DCH	RRC_DCCH_DL	radioBearerSetup	
11:33:29,325,027	RLC/MAC	AM DATA DCH	RRC_DCCH_UL	radioBearerSetupComplete	
11:33:29,373,593	RLC/MAC	AM DATA DCH	RRC_DCCH_DL	measurementControl	

ID Name	Comment or Value
1.5.1.4.3.1.5.1 dCH-Specific-FDD-Item	
1.5.1.4.3.1.5.1.1 dCH-ID	5
1.5.1.4.3.1.5.1.2 ul-TransportFormatSet	
1.5.1.4.3.1.5.1.2.1 dynamicParts	
1.5.1.4.3.1.5.1.2.1.1 sequence	
1.5.1.4.3.1.5.1.2.1.1.1 nrOfTransportBlocks	0
1.5.1.4.3.1.5.1.2.1.1.2 mode	
1.5.1.4.3.1.5.1.2.1.1.2.1 notApplicable	0
1.5.1.4.3.1.5.1.2.1.2 sequence	
1.5.1.4.3.1.5.1.2.1.2.1 nrOfTransportBlocks	1
1.5.1.4.3.1.5.1.2.1.2.2 transportBlockSize	336
1.5.1.4.3.1.5.1.2.1.2.3 mode	
1.5.1.4.3.1.5.1.2.1.2.3.1 notApplicable	0
1.5.1.4.3.1.5.1.2.1.3 sequence	
1.5.1.4.3.1.5.1.2.1.3.1 nrOfTransportBlocks	2
1.5.1.4.3.1.5.1.2.1.3.2 transportBlockSize	336
1.5.1.4.3.1.5.1.2.1.3.3 mode	
1.5.1.4.3.1.5.1.2.1.3.3.1 notApplicable	0
1.5.1.4.3.1.5.1.2.1.4 sequence	
1.5.1.4.3.1.5.1.2.1.4.1 nrOfTransportBlocks	3
1.5.1.4.3.1.5.1.2.1.4.2 transportBlockSize	336
1.5.1.4.3.1.5.1.2.1.4.3 mode	
1.5.1.4.3.1.5.1.2.1.4.3.1 notApplicable	0
1.5.1.4.3.1.5.1.2.1.5 sequence	
1.5.1.4.3.1.5.1.2.1.5.1 nrOfTransportBlocks	4
1.5.1.4.3.1.5.1.2.1.5.2 transportBlockSize	336
1.5.1.4.3.1.5.1.2.1.5.3 mode	
1.5.1.4.3.1.5.1.2.1.5.3.1 notApplicable	0
1.5.1.4.3.1.5.1.2.2 semi-staticPart	
1.5.1.4.3.1.5.1.2.2.1 transmissionTimeInterval	msec-20
1.5.1.4.3.1.5.1.2.2.2 channelCoding	turbo-coding

Figure 1.39 Transport format information in NBAP Synchronised Radio Link Reconfiguration Preparation Request

request of the core network domain based on QoS parameters received in NAS signalling from UE. SRNC may assign resources according to this requested QoS or not. It often happens that e.g. a downlink maximum bit rate of 384 kbps is requested by SGSN, but SRNC only assigns resources for 64 kbps. Unfortunately, most SRNC RANAP entities do not provide feedback to SGSN using RANAP RAB Assignment Response to signal which maximum bit rate is really possible after the initial set up of the radio bearer.

Resources assigned to the radio bearer by SRNC are always useful for root cause analysis in the case of bottleneck problems for high-speed data transmission.

In the RRC Radio Bearer Setup Request or Radio Bearer Reconfiguration message there are no maximum bit rate parameters as in RANAP messages. Instead, one will find information about RLC block size and the number of blocks in the transport block set in very abstract format. Due to this it is better to check information in the appropriate NBAP Synchronised Radio Link Reconfiguration Preparation Request message to derive information about radio bearer details.

The settings shown in Figure 1.39 for example define that on the uplink dedicated transport channel having DCH-ID = 5, the payload can be sent in the following combinations: 0 transport blocks (if no data is to be transmitted within a single transmission time interval) or one, two, three or four transport blocks, each of them 336 bits, can be transmitted in one transport format set within the time transmission interval (TTI) of 20 milliseconds.

From 3GPP 34.108 *Common Test Environments for User Equipment (UE) Conformance Testing*, which contains numerous mapping tables for transport format set – RAB definitions, it becomes evident that this particular transport format is used in the case of a 64 kbps UL RAB (see Table 1.2).

Table 1.2 Transport channel parameters for interactive or background/UL:64 kbps/PS RAB following 3GPP 34.108

Higher layer	RAB/Signalling RB	RAB
RLC	Logical channel type	DTCH
	RLC mode	AM
	Payload sizes, bits	320 (alt. 128)
	Max data rate, bps	64 000
	AMD PDU header, bits	16
MAC	MAC header, bits	0
	MAC multiplexing	N/A
Layer 1	TrCH type	DCH
	TB sizes, bits	336 (alt. 144)
	T TF0, bits	0×336 (alt. 0×144)
	F TF1, bits	1×336 (alt. 1×144)
	S TF2, bits	2×336 (alt. 3×144)
	TF3, bits	3×336 (alt. 7×144)
	TF4, bits	4×336 (alt. 10×144)
	TTI, ms	20
	Coding type	TC
	CRC, bits	16
	Max. number of bits/ TTI after channel coding	4 236 (alt. 4 812)
	Max. number of bits/radio frame before rate matching	2 118 (alt. 2 406)
	RM attribute	130 to 170

The radio bearer type for each single connection can be distinguished based on this mapping table information. However, it requires that all the important mapping tables as found in 3GPP 34.108 are implemented in performance measurement software.

1.4 STATISTICS CALCULATION AND PRESENTATION

Before KPI definitions, which include filtering procedures and aggregation definitions, can be discussed in detail it is still necessary to explain a few basics of common statistics calculations and presentation options of performance measurement data.

One may often find such requirements regarding presentation hidden in very short statements in KPI definitions or in counter/KPI names. It must be kept in mind:

A very short note in definitions for a single performance measurement can change the whole application design necessary to provide this measurement!

1.4.1 SAMPLING PERIOD

The sampling period is sometimes also called the *measurement reporting periodicity* and for many periodic measurements requested by the network from UE and Node B this reporting periodicity needs to be configured in the RNC settings.

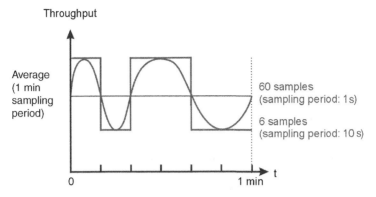

Figure 1.40   Sampling period and granularity of measurement (idealised graphs)

Looking at counters used for performance measurements it can be recognised that after the sampling period timer expires, counter values are stored in a database. Then the sampling period timer as well as counter values are reset to zero and a new sampling period starts.

Note: the smaller the sampling period the better the granularity of measurement.

The granularity period defined in 3GPP 32.401 *Performance Management (PM). Concepts and Requirements* is the period for which data already stored from e.g. RNC PM counter database is transferred to higher level performance management software.

In contrast to the granularity period defined by 3GPP, the sampling period stands for the granularity of single measurements. If, for instance, there is a one-minute data call and throughput needs to be measured this can be done using a variable sampling period (see Figure 1.40):

1. Choose sampling period = 1 minute and the analysis result will be a measurement value that gives an average throughput of the whole call.
2. Choose sampling period = 10 seconds and the analysis result will be an average measurement value for each 10 seconds of the call. This shows how throughput during the call changed from sampling period to sampling period. In a one-minute call, six different throughput samples will be computed and displayed.
3. Choose sampling period = 1 second and the analysis result will be 60 throughput samples for a one minute PS call. Better granularity allows the identification of temporary peaks (positive or negative) in single call throughput, e.g. for handover problems.

In contrast to the sampling period a turn-on-period does not have any influence on the granularity of measurement. It is used to reduce traffic load on measurement units. The problem is that if one wants to capture and analyse the traffic of a whole network it would require measurement equipment hardware that is equal to the size of the network switching equipment or even larger. Hence, a common idea to reduce load on measurement equipment is to turn on performance measurement for e.g. 5 minutes of each hour, compute sample KPI values and reports, wait 55 minutes, turn on measurement for 5 minutes again and so on.

The measurement result computed for a single sampling period can either be presented as a real value (e.g. a power level measurement from the radio interface in dB or dBm) or as an

Cylinder Capacity	0 to 1000 cc	1001 to 2000 cc	2001 to 3000 cc	3001 to 4000 cc

Figure 1.41   Bins for car categories

abstract value that displays the distribution of measurement values. These abstract values are often called bins.

1.4.2 BINS

A bin means a jar or a bottle in Japanese. What this jar or bottle contains is abstract measurement data.

No doubt that anybody who has ever completed a marketing survey has actually already worked with bins. Remember the car KPI example and imagine a fictitious survey that asks which cars people drive. It is not necessary to know the exact type and size and version of car, a general overview of the car's key performance data is enough. It could be enquired, e.g. how powerful car engines are based on their cylinder capacity, however, there may be more engine sizes on the market than one knows. Hence, the four categories shown in Figure 1.41 could have been defined in the survey.

These four categories are the bins and they will provide a good statistical overview of which cars people drive according to cylinder capacity (which can often be directly correlated to the size of car). Using these categories it is impossible to know exactly the cylinder capacity of each car, but using exact values instead of ranges for category definition would have created a thousand categories most of them containing only one match.

The advantage of having bins is to be able to visualise a fair distribution of possible values in a reasonable number of categories.

But how are bins used for performance measurement in mobile radio networks?

Throughput distribution may serve as a fairly good example. Imagine a 64 kbps RAB. If throughput samples are taken using a one-second sampling period, the momentary average throughput for each second of the call is measured. The possible maximum is a data volume of 64 kb that is sent on this RAB in one second. If one wants to have an analysis of all samples taken during the call the first step is to assign measurement samples to bins. Each bin may represent a value range of 1 kbps and the number of samples that fit into these bins/ranges will lead to a distribution as shown in Figure 1.42.

Bins are also used in 3GPP definitions of radio-related measurements (although they are not named bins). The reason for this is that integer values are more easily encoded in signalling messages than real number values. As a result 3GPP 25.133 defines a number of mapping tables such as Table 1.3 for the received total wideband power.

The number in the suffix of the reported value entry represents the integer value used in the NBAP Common Measurement Report message. The measured quantity value shows the range of measured values defined for each integer bin. The unit of measurement values is dBm.

Bin	1	2	3	...	63	64
Throughput range (kbps)	0 ... 1	1.001 ... 2	2.001 ... 3	...	62.001 ...63	63.001 ... 64
Number of samples	2	5	10	...	68	44

Figure 1.42   Throughput bins

1.4.3 THE 95TH PERCENTILE

The 95th percentile is the highest value left when the top 5% of a numerically sorted set of collected data is discarded. It is used as a measure of the peak value when one discounts a fair amount for transitory spikes.

Imagine, for example, a set of data has 100 values. After sorting numerically, discarding the top 5% means we lose the values 100/99/98/97/96, so therefore the next highest value is 95. This is the 95th percentile (Figure 1.43). The average of these values is 50.

The main reason why the 95th percentile is so useful in measuring data throughput is that it gives a very accurate picture of the cost of the bandwidth. It does not give a good impression on how fast data is transmitted on the net, but if you want to find out which technical and financial resources are necessary to provide a well-optimised theoretical throughput rate the 95th percentile will help find the answer.

For this reason billing models of Internet service providers (ISPs) use the 95th percentile to charge business customers that need constantly high bandwidth.

When PS services were introduced into mobile networks, technical specialists were hired by mobile network operators and network equipment manufacturers from the IP world (router manufacturers and ISPs). They introduced the 95th percentile into mobile network performance measurements based on their experience in IP world. However, there is a difference between mobile and fixed network IP services and it must be considered very carefully to use 95th percentile instead of averaging in each single case.

Table 1.3   Mapping table of received total wideband power from 3GPP 25.133

Reported value	Measured quantity value	Unit
RTWP_LEV_000	RTWP < −112.0	dBm
RTWP_LEV_001	−112.0 ≤ RTWP < −111.9	dBm
RTWP_LEV_002	−111.9 ≤ RTWP < −111.8	dBm
RTWP_LEV_619	−50.2 ≤ RTWP < −50.1	dBm
RTWP_LEV_620	−50.1 ≤ RTWP < −50.0	dBm
RTWP_LEV_621	−50.0 ≤ RTWP	dBm

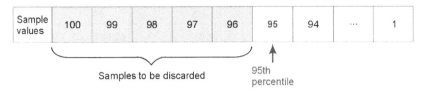

Figure 1.43   95th percentile

It is impossible to compute a 95th percentile if (rule of thumb):

$$5\% \text{ of } \sum\nolimits_{Samples} < 1 \tag{1.5}$$

In other words: the smallest number of samples that can be deleted is one sample. To compute a 95th percentile 5% of samples need to be discarded. Therefore, one sample must equal at least 5% of all samples and if one sample is 5%, the total number of samples (100%) must be 20 samples. If there are fewer than 20 samples the 95th percentile cannot be computed, because no sample can be deleted. Such a situation may especially occur in the case of bins that are pegged on the best cell if this best cell is changed very quickly. Assume that the average usage time of a radio link in soft handover is approximately 20 seconds. Within 20 seconds there are 10 bins per call computed if a two-second sampling period is used. Therefore, under such circumstances the 95th percentile cannot be computed for a single call.

1.4.4 GAUGES AND DISTRIBUTION FUNCTIONS

Gauges are mostly used for time difference measurements, e.g. the response time of an RRC radio bearer setup procedure. For a number of different calls we will get different radio bearer setup time values. From these values a mean radio bearer setup time is computed, but network operators are still interested in seeing the minimum and maximum extreme. For this reason for time difference measurements the term *min/max/mean* is often used. These three different levels are called gauges.

Since most delay and response time measurements can be fairly easy defined there is no section in this book about this topic. To identify protocol events that are related to each other please read Kreher and Ruedebusch (2005).

A basic distribution function has already been explained (see section 1.4.2 – bin distribution). Expressing that the size of bins equals the number of bin counts by using the size of a symbol (the bottles shown in Figure 1.42) is not the best way of performance measurement result visualisation. A bar diagram can be used instead and the graphical result is a diagram type also known as a distribution histogram or discrete distribution function (see Figure 1.44).

The distribution histogram is the easiest way to display a distribution function. However, some people like it to be more sophisticated and prefer to see a continuous distribution function. A subgroup of continuous distribution functions are probability distributions.

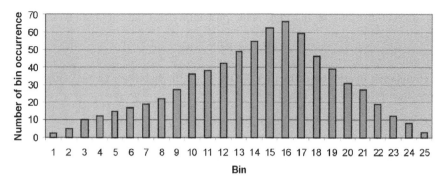

Figure 1.44   Bin distribution histogram

Probability distributions are typically defined in terms of the probability density function (PDF), but other distribution functions can be used in applications. Detailed mathematical descriptions and examples of all functions can be found in *NIST/SEMATECH e-Handbook of Statistical Methods*, http://www.itl.nist.gov/div898/handbook/. The following paragraphs about PDF and CDF are a slightly modified excerpt.

For a continuous function, the probability density function (PDF) is the probability that the variable has the value x. Since for continuous distributions the probability at a single point is zero, this is often expressed in terms of an integral between two points.

$$\int_a^b f(x)dx = Pr[a \leq x \leq b] \tag{1.6}$$

This results in the plot shown in Figure 1.45.

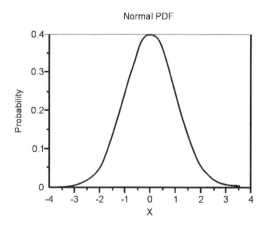

Figure 1.45   Probability distribution function (PDF)

For a discrete distribution, the PDF is the probability that the variable takes the value x.

$$f(x) = Pr[X = x] \qquad (1.7)$$

The cumulative distribution function (CDF) is the probability that the variable takes a value less than or equal to x. That is:

$$F(x) = Pr[X \leq x] = \alpha \qquad (1.8)$$

For a continuous distribution, this can be expressed mathematically as:

$$F(x) = \int_{-\infty}^{x} f(\mu)d\mu \qquad (1.9)$$

For a discrete distribution, the CDF can be expressed as:

$$F(x) = \sum_{i=0}^{x} f(i) \qquad (1.10)$$

Figure 1.46 shows a computed Continuous Distribution Function (CDF):

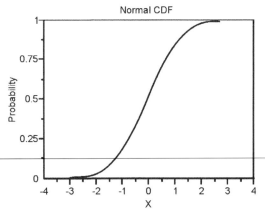

Figure 1.46   Continuous distribution function (CDF)

2

Selected UMTS Key Performance Parameters

There are three major challenges in performance measurement:

1. Defining performance measurements and KPIs.
2. Verifying measurement results.
3. Explaining measurement results.

This chapter deals with these challenges and offers solutions based on measurement scenarios and case studies. It explains how data is captured, filtered and computed, which statements can be derived from measurement analysis and which limitations apply. Furthermore, the necessary protocol knowledge to verify measurement results is presented. Once again, this is neither a perfect nor a complete overview, but a guided walkthrough. It is not a ready-to-use implementation guide, but a path to individual measurement definition and analysis.

2.1 BLOCK ERROR RATE (BLER) MEASUREMENTS

Block error rate (BLER) is an analysis of transmission errors on the radio interface. Using formal definitions BLER is a KPI, because it is a formula. It is based on analysis of cyclic redundancy check (CRC) results for radio link control (RLC) transport blocks and computed by defining the relation between the numbers of RLC transport blocks with CRC error indication and the total number of transmitted transport blocks as expressed in Equation (2.1).

$$\text{Block Error Rate (BLER)} = \frac{\sum RLC\ TransportBlocks\ with\ CRCError}{\sum RLC\ TransportBlocks} \times 100\% \quad (2.1)$$

BLER is measured separately on the uplink and downlink direction, which is mandatory, because in UTRAN frequency division duplex (FDD) mode uplink and downlink data is transmitted using different frequency bands.

There are several options for BLER filtering and computing, which will be explained in the following two sections.

2.1.1 UPLINK BLOCK ERROR RATE (UL BLER)

There is no measurement report in UTRAN that contains uplink BLER values although UL BLER is an important criterion for the radio network controller (RNC) to make handover decisions based on uplink transmission quality. Due to the fact that UL BLER is only computed and used in the RNC internally it also is only available inside RNC software and as a rule is not shown in any performance measurement statistics. Therefore, it is a typical example for performance measurement based on protocol analysis. The UL BLER is especially a very critical parameter to measure user perceived quality of services using RLC transparent mode. While AMR voice calls can compensate an UL BLER of up to 1% using AMR-specific error concealment algorithms the quality of CS videotelephony will be heavily impacted, because each block error will directly result in pixel errors in video and/or background noise in audio information. Hence, for these services UL BLER gives a correct impression of the user's perceived quality of service.

2.1.1.1 Uplink Transport Channel BLER

Usually UL BLER as well as DL BLER are computed per transport channel. This means that, e.g. for a voice call, four transport channels are used simultaneously: one dedicated channel (DCH) for the dedicated control channel (DCCH) (radio resource control (RRC) signalling) and three DCHs for voice packets (one DCH for each adaptive multi rate (AMR) A, B and C bits). This specific filtering is not mentioned in the general formula, but in 3GPP 25.215: 'The BLER estimation shall be based on evaluating the CRC of each transport block associated with the measured transport channel...'.

3GPP standards only describe BLER measurement on the downlink, but it must be assumed that the same rules apply for UL BLER computing, which is performed by the RNC internally and also computed by the performance measurement software. Therefore, for a voice call there will not be only one UL BLER, but four different UL BLER measurements. However, as explained in the next paragraphs UL BLER may appear in many different versions, but it may also be noted from above that the quote from 3GPP 25.215 is incomplete. The remainder of the quoted sentence is as follows: '... after RL combination'.

This points to another important requirement: it is necessary to take macro-diversity combining into account if the call is in a soft handover situation. A typical measurement scenario may then look as shown in Figure 2.1.

The user equipment (UE) sends UL data on three different radio links. Since each radio link is provided by a cell belonging to a different Node B, the UE is in soft handover. The same transport blocks may be sent on three different radio links, and because of the three different Node Bs involved in this scenario, also on three different Iub physical transport bearers (AAL2 SVCs). Due to high synchronisation in UTRAN all uplink transport blocks will arrive nearly simultaneously on the serving RNC (SRNC). The usual time difference between identical blocks is approximately 1 ms. Due to the fact that in a voice call transport channels carrying AMR speech information belong together, they are set up as coordinated dedicated transport channels (DCHs). Hence, transport blocks from all three AMR DCHs are found in the same UL frame protocol (FP) data frame. If the cells belong to three different Node Bs that are involved in soft handover scenario as shown in Figure 2.1, the same FP data frame is sent on each Iub interface that is involved. Following this three identical FP frames, as presented in message example 2.1, are received by SRNC.

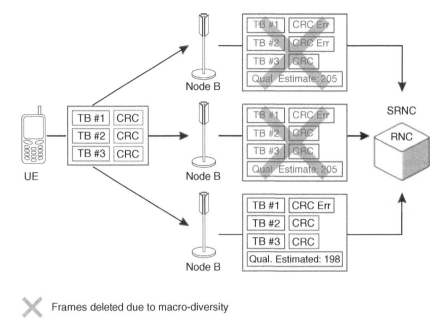

Frames deleted due to macro-diversity

Figure 2.1   UL transport blocks for UL BLER calculation

The Iub FP data frames in uplink contain some radio-related measurement results known as the quality estimate (QE). This represents the estimated bit error rate (BER) measured by Node B on the uplink radio link of a single cell. Since three cells are involved on each Iub the reported QE may be different. Following the rule explained in Chapter 1 of this book that only the UL FP data frame with the best QE is accepted by SRNC, all other frames are discarded. Although this rule is introduced to eliminate as many blok errors as possible. One erroneous transport block (#1) in the example passes the SRNC to be forwarded via IuCS and possibly core network interfaces to the B-party of the call.

Note 1: if UE is located in cell border areas CRC errors occur more often than usual. However, in cell border areas soft handover procedures are triggered, too. To measure UL BLER correctly in such situations it is first necessary to see the first uplink data frame transmitted on the new Iub interface as a trigger for the start of macro-diversity combining of uplink frames. It is also useful to correlate BLER measurements with the active set size of calls to find out which specific impact softer and soft handover situations have on radio transmission quality.

Note 2: some network equipment manufacturers do not use QE for the macro-diversity combining algorithm. Instead they select the best frame using another uplink radio quality parameter associated with the receiving radio link, e.g. uplink signal-to-interference ratio (SIR).

The size and type of each transport block is indicated by the transport format index, a value that corresponds with the settings of the transport format set as seen in Node B application part (NBAP) radio link setup or radio link reconfiguration preparation procedures. If the transport block is erroneous or not is indicated by a so-called CRC indicator associated with each transport block. This CRC indicator is a parameter in an FP trailer, an appendix transmitted on Iub in the uplink direction only. It indicates the result of a CRC after

Message example 2.1 UL FP data frame including transport blocks and CRC indicators

I TS 25.322 (RLC) / 25.321 (MAC) / 25.435, 25.427 (FP) - V3.13.0 (RLC/MAC) FP DATA DCH (= FP Data Frame DCH)	
I FP Data Frame DCH	
I FP: VPI/VCI/CID	I "188/65/234"
I	
I FP: Radio Mode	I FDD (Frequency Division Duplex)
I FP: Direction	I Uplink
I FP: Transport Channel Type	I DCH (Dedicated Channel)
I 1 FP: Header	
I 1.1 FP: Header CRC	I '35'H
I 1.2 FP: Frame Type	I Data
I 1.3 FP: Connection Frame Number	I 220
I 1.4 FP: Spare	I 0
I 1.5 FP: Transport Format Index	I 2
I 1.6 FP: Spare	I 0
I 1.7 FP: Transport Format Index	I 1
I 1.8 FP: Spare	I 0
I 1.9 FP: Transport Format Index	I 1
I 2 Transport Block Set DCH	
I 2.1 FP: DCH Index	I 0
I 2.2 FP: Transport Block	
I 2.2.1 MAC: Target Channel Type	I DTCH (Dedicated Traffic Channel)
I 2.2.2 MAC: RLC Mode	I Transparent Mode
I 2.2.3 RLC: Whole Data	I '1010001101010001001011011001'B
I 3 Transport Block Set DCH	
I 3.1 FP: DCH Index	I 1
I 3.2 FP: Transport Block	
I 3.2.1 MAC: Target Channel Type	I DTCH (Dedicated Traffic Channel)
I 3.2.2 MAC: RLC Mode	I Transparent Mode
I 3.2.3 RLC: Whole Data	I '01010010000011011100111010101110'B
I 4 Transport Block Set DCH	
I 4.1 FP: DCH Index	I 2
I 4.2 FP: Transport Block	
I 4.2.1 MAC: Target Channel Type	I DTCH (Dedicated Traffic Channel)
I 4.2.2 MAC: RLC Mode	I Transparent Mode
I 4.2.3 RLC: Whole Data	I '01000100110100000011001010010110001'B
I 5 FP: Trailer	
I 5.1 FP: Quality Estimate	I 132
I 5.2 FP: CRC Indicator (Transport Block)	I Not Correct
I 5.3 FP: CRC Indicator (Transport Block)	I Correct
I 5.4 FP: CRC Indicator (Transport Block)	I Correct
I 5.5 FP: Padding	I 00

transmission of the transport block on the radio interface. However, not every transport channel is protected by CRC. If CRC is activated or not and how many bits are used in the check sequence is indicated in NBAP Radio Link Setup or Radio Link Reconfiguration Request messages. In downlink there is no BLER measurement on channels that are not CRC protected, but for uplink data frames, the FP entity of Node B considers blocks as transmitted correctly if no CRC is executed. If the target is to measure UL BLER with highest precision, the BLER measurement application must check if CRC is defined for each transport channel set up by NBAP. Otherwise CRC indicator values reported for this transport channel in frame protocol will be ignored for UL BLER measurement.

*Note: UL transport channel BLER is usually not measured on the random access channel (RACH) although it is theoretically possible. This is to be aligned with 3GPP 25.215 that says: 'The measurement **Transport channel BLER** does not apply to transport channels mapped on a P-CCPCH and a S-CCPCH'. Transport channels mapped on mentioned common physical channels are the broadcast channel (BCH) and forward access channels (FACHs). If BLER is not measured on FACHs (downlink) a corresponding measurement on RACHs (uplink) does not seem to make much sense.*

2.1.1.2 UL BLER per Call

UL BLER can also be used to estimate the uplink transmission quality of a call. In this case it is not necessary to differentiate between transport channels that carry user plane and control plane information. Merely count transport blocks and their CRC indicators according to the standard formula given at the beginning of this chapter, no matter to which transport channel single transport blocks belong.

2.1.1.3 UL BLER per Call Type

Another possibility is to map UL BLER to the type of call: voice, packet switched (PS) data, video-telephony, multi-RAB, signalling. Depending on differentiation based on higher layer information more types of call may be defined according to end-user services. However, definitions should be considered carefully. Transmitting a short message in uplink direction only requires a few transport blocks. In addition it must be considered that quality of end-user services like speech can only be measured on user plane transport channels. Especially for voice calls it is further necessary to define filter functions or special algorithms if there is no CRC on transport channels that carry AMR B and C class bits.

2.1.2 DOWNLINK BLOCK ERROR RATE (DL BLER)

Compared to uplink block error rate DL BLER does not need to be computed by any performance measurement equipment based on transport block counters. Neither it is meaningful to compute it on Iub, because transmission errors will appear on Uu – **after** data has been sent to the cell via Iub in downlink.

The job to compute and report DL BLER is assigned to UE using the RRC measurement control message (message example 2.2).

The measurement can be reported periodically or if a predefined threshold of CRC errors on downlink transport channels is exceeded. For a threshold event triggered reporting using event 5A will be monitored. From the performance monitoring point of view the disadvantage of event-triggered reporting is that one can no longer follow a certain measurement over a long period of time. Instead only single peak values (exceptions) are reported while all normal values are hidden to lower load on SRNC caused by measurement tasks. In the case of DL BLER measurements presented in Figure 2.2 only the extreme values shown on the right side of the histogram have a chance of being reported.

Message example 2.2 RRC measurement control to set up periodical DL BLER reporting

ID Name	Comment or Value	
TS 25.331 DCCH-DL - V5.9.0 (RRC_DCCH_DL) measurementControl (= measurementControl)		
dL-DCCH-Message		
2 message		
2.1 measurementControl		
2.1.1 r3		
2.1.1.1 measurementControl-r3		
2.1.1.1.1 rrc-TransactionIdentifier	0	
2.1.1.1.2 measurementIdentity	16	
2.1.1.1.3 measurementCommand		
2.1.1.1.3.1 setup		
2.1.1.1.3.1.1 qualityMeasurement		
2.1.1.1.3.1.1.1 qualityReportingQuantity		
2.1.1.1.3.1.1.1.1 dl-TransChBLER	**true**	
2.1.1.1.3.1.1.2.1.1 reportingAmount	ra-Infinity	
2.1.1.1.3.1.1.2.1.2 reportingInterval	ri11	
2.1.1.1.4 measurementReportingMode		
2.1.1.1.4.1 measurementReportTransferMode	acknowledgedModeRLC	
2.1.1.1.4.2 **periodicalOrEventTrigger**	**periodical**	

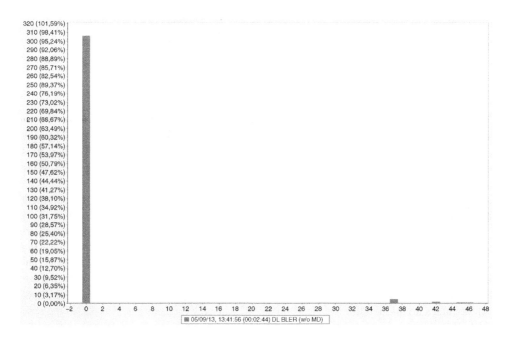

Figure 2.2   Distribution of DL BLER measurement reports monitored during a single call

Message example 2.3 RRC Measurement report containing DL BLER report for a speech call

ID Name	Comment or Value	
TS 25.331 DCCH-UL - V5.9.0 (RRC_DCCH_UL) measurementReport (= measurementReport)		
uL-DCCH-Message		
2 message		
2.1 measurementReport		
2.1.1 measurementIdentity	16	
2.1.2 measuredResults		
2.1.2.1 qualityMeasuredResults		
2.1.2.1.1 blerMeasurementResultsList		
2.1.2.1.1.1 bLER-MeasurementResults		
2.1.2.1.1.1.1 **transportChannelIdentity**	32	
2.1.2.1.1.1.2 **dl-TransportChannelBLER**	0	
2.1.2.1.1.2 bLER-MeasurementResults		
2.1.2.1.1.2.1 **transportChannelIdentity**	1	
2.1.2.1.1.2.2 **dl-TransportChannelBLER**	0	
2.1.2.1.1.3 bLER-MeasurementResults		
2.1.2.1.1.3.1 **transportChannelIdentity**	2	
2.1.2.1.1.4 bLER-MeasurementResults		
2.1.2.1.1.4.1 **transportChannelIdentity**	3	
2.1.2.1.2 modeSpecificInfo		
2.1.2.1.2.1 fdd	0	

DL BLER values for each transport channel of a connection are reported by the UE using the RRC measurement report message as shown in message example 2.3.

In this example the UE computed the transport channel BLER for downlink transport channels. DL BLER is only measured on those channels for which a CRC procedure has been defined during channel establishment. Whether a CRC for a certain transport channel applies can be ascertained from transport format settings found in the NBAP Radio Link Setup Request or NBAP Radio Link Reconfiguration Preparation Request message.

Report values are sent in bin format. Conversion from percentage ratio into bins follows the mapping table found in 3GPP 25.133 (see Table 2.1).

There is no unit in this measurement, but it could be expressed in %. Bin value 62 (BLER_LOG_62), for example, represents a BLER in the range from 74% to 86%. If the

Table 2.1 Bin mapping table for DL transport channel BLER reports

Reported value	Measured quantity value
BLER_LOG _00	Transport channel BLER $= 0$
BLER_LOG _01	$-\infty < \text{Log}_{10}(\text{transport channel BLER}) < -4.03$
BLER_LOG _02	$-4.03 \leq \text{Log}_{10}(\text{transport channel BLER}) < -3.965$
BLER_LOG _03	$-3.965 \leq \text{Log}_{10}(\text{transport channel BLER}) < -3.9$
...	...
BLER_LOG _61	$-0.195 \leq \text{Log}_{10}(\text{transport channel BLER}) < -0.13$
BLER_LOG _62	$-0.13 \leq \text{Log}_{10}(\text{transport channel BLER}) < -0.065$
BLER_LOG _63	$-0.065 \leq \text{Log}_{10}(\text{transport channel BLER}) \leq 0$

reported bin value is displayed as a percentage value, it is necessary to specify which value to show when the UE sends reports with DL BLER = 62. Will the lowest assumed value of 74%, the highest possible value of 86% or the value in the middle of the range (80%) be displayed and used for further calculations? Whatever is decided will have a significant impact on calculated average values and graphical presentation.

Note: it is not possible to derive exact DL BLER measurement results in % from values found in the RRC measurement report.

2.1.2.1 DL BLER per Call or Service

It is not easy to compute DL BLER for a certain call or call type. The most important reason is the bin format used in the RRC measurement report. In the case of a voice call there are four or two (if no CRC is activated for transport channels that carry AMR B and C bits) different bins reported by the UE to the SRNC, which makes it difficult to build an average on bin values. This is explained in the following example.

Imagine a conversion function for ratio/bins based on a logarithmic function as described in the imaginary mapping table shown in Table 2.2.
There are two ratios measured and converted into bins:

$$\text{Measure } \#1 = 10^{-2} \Rightarrow \text{bin } 20$$
$$\text{Measure } \#2 = 10^{-3} \Rightarrow \text{bin } 30$$

The *correct average* of these measurement results is:

$$\frac{(10^{-2} + 10^{-3})}{2} = 0.0055 = 10^{-2.25936} \Rightarrow \text{bin } 22$$

This calculation proves that to calculate the average of the bin values in the case of the above measurement leads to a *wrong result*, because:

$$\frac{(20 + 30)}{2} = 25$$

And the average measurement result is not represented by bin 25!

It must also be known which bearer service applies to which transport channel. For a voice call there are four transport channels: one for the signalling radio bearer (RRC signalling) and three for AMR speech. While monitoring the speech service BLER it must be known on which channel AMR A-class bits are transmitted. The problem is that especially in the case

Table 2.2 Imaginary bin mapping table to demonstrate calculation of bin average

Ratio value	Bin value
10^{-2} to $10^{-2.1}$	20
$10^{-2.1}$ to $10^{-2.2}$	21
. . .	. . .
10^{-3} to $10^{-3.1}$	30

of multi-RAB calls it cannot be distinguished from transport channel ID on which channel A-class bits are transmitted, because CS and PS services of other calls may use transport channels with the same ID. Hence, the individual transport format settings for each transport channel have to be analysed based on NBAP signalling information and stored as call-specific context as long as the call is active. Looking at this context is could be evaluated by performance measurement software at any time on which transport channel specific speech information (e.g. AMR A-class bits) or IP data is transmitted. Although it would not be impossible to implement this in performance measurement software the problem is that in a live network environment with hundreds or thousands of calls to be analysed simultaneously, there is a lot of information that the software needs 'to keep in mind', which leads to a significant impact on system performance. For this reason this feature might not be found in existing applications.

2.1.3 CORRELATION OF BLER AND OTHER MEASUREMENTS

It can be observed that the number of monitored CRC errors increases especially a short time before and short time after radio link additions in soft handover situations. Most errors, which cannot be compensated using macro-diversity filtering, also occur in these time frames, which are usually not longer than 600 ms (maximum), and on average are approximately 250 ms. Therefore, it makes sense to correlate RRC measurement reports as well as messages used to execute the RRC Active Set Update procedure with BLER measurements (see Figure 2.27), because it helps to optimise settings for measurements that trigger radio link additions. Similar investigation could be done for triggering hard handover to different UTRAN frequencies or to different radio access technologies (RATs).

Another possible correlation can be found when analysing UL BLER and throughput of PS calls as illustrated in the time evaluation diagram in Figure 2.3.

Looking at this figure it is evident that UL BLER is always high if throughput is low and vice versa. At first sight it looks as if uplink transport channel throughput is low due to high UL BLER, but this conclusion is wrong. If UL BLER is high for PS services the receiving RLC entity of the SRNC will order a retransmission of corrupted frames and for retransmissions the full bandwidth of a 64 kbps radio bearer is available without limitation. Hence, high UL BLER cannot be the root cause of low transport channel throughput. It must therefore be assumed that throughput is low because there is no more data to be transmitted in the uplink direction, and due to radio transmission conditions UL BLER always rises if any data is transmitted in the uplink direction.

Another correlation is probably more meaningful: to compare UL BLER with SIR error reported by Node B for connections in cell 512. SIR error is sent by Node B if the base station is no longer able to adjust the uplink transmission power of an UE by sending transmit power commands (TPC) in downlink. Actually SIR should have the same value for all users in a cell, but if radio transmission conditions on uplink frequency change quickly, the SIR target cannot be maintained and following this Node B sends an SIR error report to the RNC that may define a new SIR target or make a handover decision. A common reason for frequently sent SIR errors is that the UE is moving quickly, e.g. because it is located in a vehicle. Movements can be correlated to handover triggers, high UL BLER can be correlated to handover triggers. Thus, it makes sense to correlate SIR error with UL BLER measurements.

Figure 2.4 proves that UL BLER significantly rises if a higher SIR error (positive or negative difference from 0 dB) is reported. This measurement correlation by itself does

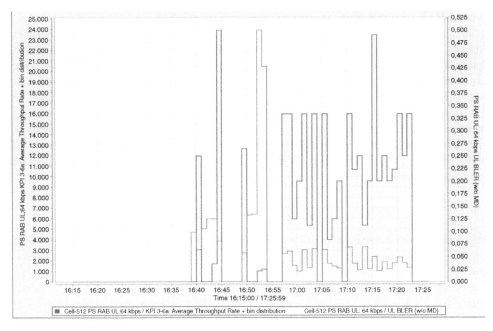

Figure 2.3 UL BLER correlated with UL transport channel throughput

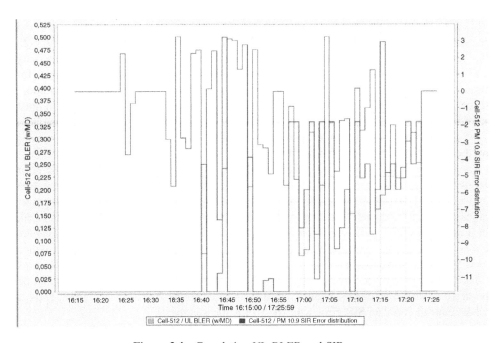

Figure 2.4 Correlation UL BLER and SIR error

not indicate any error. It just highlights once again how much UL BLER depends on the mobility of the UE and changing radio transmission conditions on the radio interface. Another useful approach is to show BLER measurement results correlated with active set size of calls as shown in figure 2.27.

2.2 RADIO-RELATED MEASUREMENTS

Having introduced some radio-related measurements in the section about BLER it is now necessary to give a deeper insight into this topic. First an overview on which protocol layers such measurements can be found will be given followed by a short description of parameters and related protocol procedures. In general, all radio-related measurements are well defined and described in 3GPP 25.215 and 25.133. Therefore, this section will provide a summary combined with information on message flows and some measurement result examples. All descriptions given follow 3GPP Release 5 standards version 5.7.0 if not otherwise stated.

3GPP 25.215 *Physical Layer – Measurements (FDD)* defines two major groups: UE measurement abilities and UTRAN measurement abilities. Looking at protocol functions it emerges that all reports related to UE measurement abilities are provided by the RRC protocol while NBAP and Iub FP are used for reporting values of UTRAN measurement abilities. A closer insight to NBAP reveals that there are two different categories defined: dedicated measurement and common measurement procedures. While common measurement procedures are related to cells, dedicated procedures are related to dedicated physical channels connecting UEs and network.

2.2.1 RADIO LINK QUALITY PARAMETERS AND FLOW CONTROL IN LUB FRAME PROTOCOL (FP)

The most important radio link quality parameters found in FP frames were introduced in the section about BLER and in the general description of the macro-diversity combining procedure. These are the CRC indicators and quality estimate, both found in the so-called FP trailer, an appendix sent in the uplink direction to the SRNC added by Node B to every RLC transport block set (see Figure 2.5).

The function of CRC indicators has already been described in section 2.1.1. In contrast to block error rate (BLER), quality estimate (QE) is the bin-decoded bit error rate (BER) measured on the uplink physical channel. There are two options regarding the type of channel: QE is equivalent either to the physical channel BER or the transport channel BER (see Figure 2.6). The difference is that the transport channel BER is the remaining BER after convolutional decoding. Convolutional coding is used to provide error detection and error

```
3 FP:  Trailer
10010010   3.1 FP:   Quality Estimate                          146
0-------   3.2 FP:   CRC Indicator (Transport Block)           Correct
-0------   3.3 FP:   CRC Indicator (Transport Block)           Correct
--000000   3.4 FP:   Padding                                   0
***B2***   3.5 FP:   Payload CRC                               6ac3
```

Figure 2.5   FP trailer with CRC indicator and quality estimate

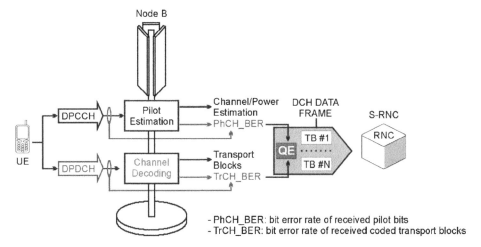

Figure 2.6   Different sources of QE/uplink BER

correction mechanisms for each transport channel. Each transport block that is sent on a defined transport channel is encoded by the convolutional coder of the sender and decoded by the receiver. Higher coding rates provide higher protection against bit transmission errors on the radio interface. Common coding rate on dedicated transport channels in UTRAN is 1/2 or 1/3, which means that 2 or 3 bits are sent on the radio interface to transmit 1 bit of real information. In general it is correct to say that protection against bit errors is achieved by adding redundant information to data streams. The transport channel BER expresses how well convolutional coding error protection has been able to minimise the number of bit errors in a transmitted RLC block set while the physical channel BER is the bit error rate measured on the pilot sequences of the dedicated physical control channel (DPCCH). Transport channels are sent on the parallel dedicated physical data channel (DPDCH) in the uplink direction. If the QE-selector information element in the NBAP Radio Link Setup Request or NBAP Radio Link Radio Reconfiguration Preparation Request message is set to 'selected' then QE is the transport channel BER. If QE-selector has value 'non-selected' this means that no transport channel BER is available and QE shows the physical channel BER.

The QE/BER is encoded in 256 bins. Mapping between bins and ratio values is as shown in Table 2.3.

Table 2.3   Bin mapping table BER/QE (from 3GPP 25.133)

Reported value	Measured quantity value
PhCh_BER_LOG_000	Physical channel BER $= 0$
PhCh_BER_LOG_001	$-\infty < \mathrm{Log}_{10}(\text{physical channel BER}) < -2.06375$
PhCh_BER_LOG_002	$-2.06375 \leq \mathrm{Log}_{10}(\text{physical channel BER}) < -2.055625$
PhCh_BER_LOG_003	$-2.055625 \leq \mathrm{Log}_{10}(\text{physical channel BER}) < -2.0475$
$\cdots$	$\cdots$
PhCh_BER_LOG_253	$-0.024375 \leq \mathrm{Log}_{10}(\text{physical channel BER}) < -0.01625$
PhCh_BER_LOG_254	$-0.01625 \leq \mathrm{Log}_{10}(\text{physical channel BER}) < -0.008125$
PhCh_BER_LOG_255	$-0.008125 \leq \mathrm{Log}_{10}(\text{physical channel BER}) \leq 0$

The same CRC indicator bit as used for transport blocks sent on the DCH can also be found on the random access channel (RACH). Its definition and meaning are the same as for the CRCI on the DCH.

There is another interesting measurement defined on common transport channels. On the RACH and common packet channel (CPCH) an access preamble mechanism is used to avoid collisions. During these preamble phases of the transmission Node B can determine the propagation delay between UE and Node B antennas. The calculated value is included in the RACH DATA FRAME and CPCH DATA FRAME. The propagation delay is given in terms of chip periods. The value range of propagation delay is 0 to 765 chips with a granularity of 3 chips. The field is coded as an unsigned integer value as follows: $1 = 3$ chips, $2 = 6$ chips, and so on.

On the downlink FACH the power used for sending frames on this common transport channel can be adjusted for single frames if there is transmit power level information embedded in the downlink FP frame on the FACH. Transmit power level indicates the required downlink power as negative offset to the configured maximum power of the associated secondary common control physical channel (S-CCPCH – this is the physical channel that carries the FACH). The offset can be 0 to 25.5 dB with a step size of 0.1 dB. The value is coded as an unsigned integer as follows: $1 = 0.1$ dB, $2 = 0.2$ dB, and so on.

FP control frames sent in downlink can contain information necessary to perform inner loop power control. The SRNC sends an FP Outer Loop Power Control message to adjust the uplink SIR target for the inner power loop power control mechanism on the radio interface. The wording sounds strange, because the FP message is part of the outer loop power control procedure while the included parameter is used in inner loop power control.

Figure 2.7 explains how outer and inner loop power control mechanisms work. Node B receives an SIR target calculated by the SRNC. Using transmit power commands (TPCs),

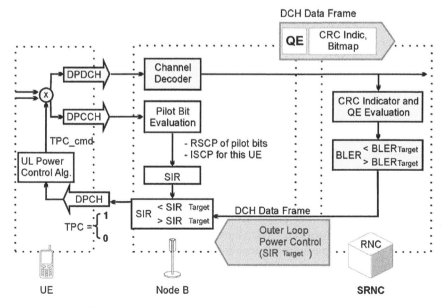

Figure 2.7 Outer and inner loop power control overview

sent 1500 times per second on downlink dedicated physical channel (DPCH), Node B tries to control the uplink Tx power of UE in a way that means there is no difference to the previously set SIR target. TPCs consist of a single control bit. This bit set to '1' means that UE will increase Tx power, '0' means UE will decrease power before sending the next uplink data block. In soft handover scenarios 1500 TPCs per second are received on the UE side from every cell involved in the active set. The UE needs to combine all TPC signals from different cells and compute a summary TPC command (TPC_cmd). The TPC_cmd can have the following values: $+1$ (increase UL power), -1 (decrease UL power) and 0 (keep UL power).

Using CRC indicators included in uplink FP data frames the SRNC computes UL BLER and compares this value internally with the UL BLER target, which is also used internally to calculate the SIR target. In addition, it is checked if the QE/BER is in a predefined range. Node B calculates SIR for each active connection and optionally reports these values using NBAP dedicated measurement reports (please read section dealing with dedicated NBAP measurements to learn how SIR is computed).

If the SIR target needs to be adjusted an FP Outer Loop Power Control message is sent containing the new SIR target. Transmission of SIR error reports using NBAP procedures provides additional protection against fast fading effects.

These are all radio-related frame protocol procedures found in the so-called R99 cells: cells that do not support high-speed downlink packet access (HSDPA). In an HSDPA-capable R5 cell, in addition to the described FP, a special HS-DSCH (high speed downlink shared channel) frame protocol is running, which provides some flow control mechanisms for HS-DSCH defined in 3GPP 25.877 *High Speed Downlink Packet Access: Iub/Iur Protocol Aspects*. It might be interesting to monitor these capacity allocation procedures and compare their settings with throughput measurements on the HD-DSCH.

Note: despite the protocol name HS-DSCH FP (D = 'downlink') the HS-DSCH Capacity Allocation Response message is sent in the uplink direction. Actually there is no reason to be confused, because on AAL2 SVC for FACH frame protocol Uplink Synchronisation control frames are also sent. The difference is that FP control frames are signalling messages exchanged bidirectionally between FP entities while FP data frames transport RLC transport block sets across the Iub, which in the case of FACH and HS-DSCH are of course only sent in the downlink direction via the radio interface.

2.2.2 NBAP COMMON MEASUREMENTS

NBAP common measurement procedures are either related to a single cell or to a defined common transport channel within a single cell. Hence, the required aggregation level of these measurements is always *per cell*.

NBAP common measurement reports containing values of received total wideband power (RTWP) and transmitted carrier power have been monitored in live networks since the earliest phases of deployment. The number of acknowledged physical random access channel (PRACH) preambles is monitored more often. Detected and acknowledged preambles on the CPCH are also defined by 3GPP, but usually the CPCH is not used in current network installations and hence this measurement is not monitored.

All NBAP common measurement procedures are initialised in the same way using the NBAP common measurement initialisation procedure.

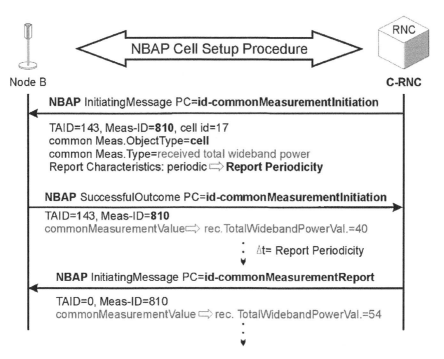

Figure 2.8 Initialisation and periodic reporting of received total wideband power

Figure 2.8 shows how this is done in particular for received total wideband power. In R99 cells this measurement is usually set up after a successful cell setup procedure. The NBAP Common Measurement Initialisation Request message is the only message that contains information about the correlation of measurement ID (Meas-ID) and cell ID (C-ID). Once measurement is started only the measurement ID associated with the cell is used in NBAP measurement report messages. This fact makes it hard for the topology module to detect to which cell a certain measurement value belongs. This statement is true for all NBAP common measurements.

However, there is an exception in the case of received total wideband power, because this measurement is not only reported using NBAP Common Measurement Report messages, but also in Successful Outcome messages of NBAP radio link setup and radio link addition procedure measurement. Since the Initiating Message of radio link setup or radio link addition contains the cell ID as a mandatory parameter correlation to cell ID in these cases is easy, RTWP measurement values reported using NBAP common measurement report still cannot be correlated to cell ID. The best possible dimension filtering looks as follows.

The mandatory job for a topology detection module is to identify all measurement IDs belonging to a single Node B. Imagine a Node B having three cells and for each cell a common measurement procedure for transmitted carrier power and received total wideband power is initialised. Based on this filtering from measurement IDs average values of measurement values related to all six measurement IDs valid for this Node B are shown in Table 2.4. An additional column shows the RTWP average computed from measurement values found

Table 2.4 Presentation of received total wideband power and transmitted carrier power measurements if correlation cell ID to Measurement ID unknown

	Transmitted carrier power	RTWP (common measurement report)	RTWP (RL setup/addition)
Node B 1			
Meas-ID 1	Session average		
Meas-ID 2	Session average		
Meas-ID 3	Session average		
Meas-ID 4		Session average	
Meas-ID 5		Session average	
Meas-ID 6		Session average	
C-ID 100			Session average
C-ID 200			Session average
C-ID 300			Session average

in the NBAP Radio Link Setup/Radio Link Addition Response message, which are correlated to cell IDs belonging to this Node B. Two columns necessary for RTWP as long as the correlation C-ID/Meas-ID is unknown.

Special dialogue windows in the graphic user interface of the software allow manual mapping of Meas-ID and C-ID. Such a correlation can be assumed from measurement results for RTWP, because RTWP values found in NBAP common measurement reports and Radio Link Setup/Addditon Response messages must have similar values.

The advantage of this presentation format in general is that despite missed correlation to cell the software user will have a good impression of what happens in the single cells of Node B. If abnormal behaviour is observed in a single measurement and troubleshooting becomes necessary, a look at the RNC settings (found in OMC software) will help to identify to which particular cell a measurement ID belongs. This provides a choice of one out of six using the format shown in Table 2.4. It is typically a choice of one out of more than 1000 for those formats that do not provide intelligent filter and correlation functions.

Once it is known which Meas-ID belongs to which C-ID, the table formatting needs to be changed so that RTWP measurements are not shown per cell and measurement values of previously separate columns are merged. An average value of RTWP per Meas-ID can still be shown optionally (see Table 2.5).

It still needs to be discussed in the following sections which technical statements can be made based on measured values of NBAP common measurement procedures and how these measurements values can be correlated to other performance parameters.

2.2.2.1 Transmitted Carrier Power

Transmitted carrier power is the ratio between the current transmission power used to transmit all downlink traffic of a single cell and the maximum configured transmission power for this cell.

In other words transmitted carrier power gives feedback information about how much of the totally available downlink transmission power is currently used and how many reserves are left.

Table 2.5 Presentation of received total wideband power and transmitted carrier power measurements for known correlation cell ID to Meas-ID

	Transmitted carrier power	RTWP (common measurement report)	RTWP (RL setup/addition)
Node B 1			
Meas-ID 1	Session average		
Meas-ID 2	Session average		
Meas-ID 3	Session average		
Meas-ID 4		Session average	
Meas-ID 5		Session average	
Meas-ID 6		Session average	
C-ID 100	Session average	Session average	
C-ID 200	Session average	Session average	
C-ID 300	Session average	Session average	

The transmitted carrier power value is encoded in bins that represent percentage values between 0 and 100 (see Table 2.6).

2.2.2.2 NBAP Common Measurement Enhancements in Release 5

Additional common measurements introduced in Release 5 standards are *transmitted carrier power of all codes not used for HS-PDSCH or HS-SCCH transmission, HS-DSCH provided bit rate* and *HS-DSCH required power*. The first of these measurements defines transmitted carrier power excluding all power used for HSDPA channels in HSDPA capable cells. As a rule it is used instead of the Release 99 transmitted carrier power measurement if the cell is HSDPA capable. The mapping table for bin encoding of measurement values is the same as shown in Table 2.6. The HS-DSCH provided bit rate is reported as an integer value that represents the DL transport channel throughput on HS-DSCH in bps. The HS-DSCH required power is an integer value in the range from 0 to 1000 that stands for parts per thousands (ppt) of maximum transmission power.

Another interesting option was introduced by the implementation of Release 5 NBAP protocol version although it was already defined in Release 99. This option is the event-triggered reporting of NBAP measurements. The target of this feature is to minimise the

Table 2.6 Bin mapping table transmitted carrier power (from 3GPP 25.133)

Reported value	Measured quantity value	Unit
UTRAN_TX_POWER _000	Transmitted carrier power $= 0$	%
UTRAN_TX_POWER _001	$0 <$ Transmitted carrier power ≤ 1	%
UTRAN_TX_POWER _002	$1 <$ Transmitted carrier power ≤ 2	%
UTRAN_TX_POWER _003	$2 <$ Transmitted carrier power ≤ 3	%
. . .	. . .	. . .
UTRAN_TX_POWER _098	$97 <$ Transmitted carrier power ≤ 98	%
UTRAN_TX_POWER _099	$98 <$ Transmitted carrier power ≤ 99	%
UTRAN_TX_POWER _100	$99 <$ Transmitted carrier power ≤ 100	%

Table 2.7 Definitions and meanings of event-triggered NBAP reporting

Report characteristics settings	Meaning
On demand	Node B returns current measurement result value immediately
Periodic	Node B reports measurement values periodically (as defined in Release 99 and shown in Figure 2.9)
Event A	Node B sends measurement report if measurement result is **above** a requested threshold
Event B	Node B sends measurement report if measurement result is **below** a requested threshold
Event C	Node B sends measurement report if measurement result **rises by an amount greater than** a requested threshold
Event D	Node B sends measurement report if measurement result **falls by an amount greater than** a requested threshold
Event E	Node B starts periodic measurement reporting if measurement value **rises** above a requested threshold and stops periodic reporting if measurement value **falls** below the same or different threshold
Event F	Node B starts periodic measurement reporting if measurement value **falls** below a requested threshold and stops periodic reporting if measurement value **rises** above the same or different threshold

number of NBAP measurement reports sent on the Iub interface and in turn to minimise the measurement load on CRNC software. All defined reporting events are valid for both NBAP common and NBAP dedicated measurements. Event settings and appropriate parameters are found in the report characteristics information element of the NBAP Common/ Dedicated Measurement Initialization Request message. Report characteristics have the main values/meanings explained in Table 2.7.

Figure 2.9 shows a typical call flow and measurement graph example for an RTWP measurement that starts periodical reporting triggered by event E.

2.2.2.3 Received Total Wideband Power

Received total wideband power is the total power of all signals received in the uplink frequency band on the cell antenna, no matter if these signals are uplink physical channels sent by UE or interference from sources outside the UTRAN.

It is also correct to say that received total wideband power is the total noise received at the cell antenna on the uplink frequency, because on this frequency each UE is an interferer of all other UEs in the cell and in addition, interference from other signal sources is also measured. Such external interference is any high-frequency signal that interferes with UTRAN uplink frequencies, e.g. sidebands of radio equipment working on different frequencies than UTRAN or interference caused by electronic devices (see Figure 1.7).

In general received total wideband power represents the uplink load in a UTRAN cell as shown in Figure 2.10. For each connection requested to be set up, admission control in the SRNC calculates the expected additional load for the cell. As long as there is no risk of exceeding the thresholds $P_{Rx\ Target}$ and $P_{Rx\ Overload}$ requested connections will be granted. If

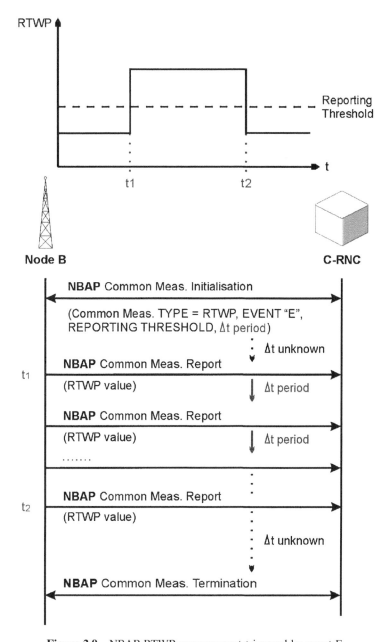

Figure 2.9   NBAP RTWP measurement triggered by event E

$P_{Rx\ Target}$ is reached or exceeded new UEs will not be allowed to set up a radio link in this cell, e.g. an RRC Connection Request is answered by the SRNC with RRC Connection Reject. Optionally rejected UEs may be redirected to other cells in the UTRAN or a different RAT. If this rerdirection function is supported by the network equipment it is meaningful to

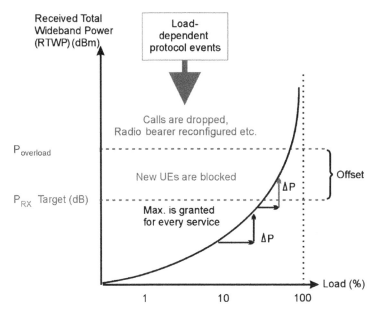

Figure 2.10   Received total wideband power and uplink cell load

calculate an RRC redirection success/failure rate in addition to the RRC blocking rate discussed in section 2.11.1.1.

If $P_{Rx\ overload}$ is exceeded, the admission control and packet scheduler in the SRNC will react with reconfiguration and reallocation of dedicated cell resources: decreasing data transmission rates for PS services, send UEs with active PS background calls to CELL_FACH state by performing intra-cell handover (channel-type switching) and also network-controlled drops of single calls will be monitored. If the RTWP exceeds the $P_{Rx\ overload}$ threshold this is also an indicator that the cell range is shrinking due to so-called *cell breathing*.

There is no doubt that a high RTWP has a large impact on the user-perceived quality of service and correlation to UTRAN KPIs like RRC blocking rates and call drop rates makes a lot of sense as suggested in Figure 2.11.

If RRC blocking rates and/or call drop rates in a defined cell area are rising the network operator can react by defining redirection procedures or by increasing the number of cells in the relevant area. For additional cells additional radio resources (codes, frequencies) are necessary.

Received total wideband power values are encoded in 622 bins, each bin represents a range of 0.1 dBm as explained in Table 2.8.

Definition: dBm is an abbreviation for the power ratio in dB (decibel) of the measured power referenced to one milliwatt.

The term dB is mainly used for an attenuation or an amplification, but dBm is used for a measured power. Zero dBm equals one milliwatt. A 3 dBm increase represents roughly doubling the power, which means that 3 dBm equals 2 milliwatt.

To read more details about other NBAP common measurements such as acknowledged PRACH preambles please refer to 3GPP 25.215 *Physical Layers – Measurements (FDD)* and 3GPP 25.133 *Requirements for Support of Radio Resource Management (FDD)*.

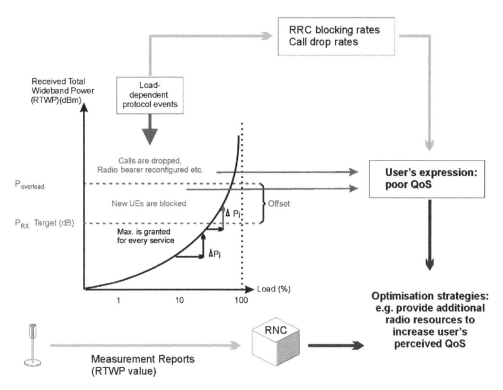

Figure 2.11   Correlation received total wideband power and UTRAN KPIs

2.2.3 NBAP DEDICATED MEASUREMENTS

Dedicated measurements are bound to the existence of dedicated channels. The UE must be in CELL_DCH state to have dedicated channels in use. When dedicated channels are set up, e.g. during the RRC connection setup procedure, one or more NBAP dedicated measurements are initialised.

Table 2.8   Bin mapping table of received total wideband power (from 3GPP 25.133)

Reported value	Measured quantity value	Unit
RTWP_LEV _000	$RTWP < -112.0$	dBm
RTWP_LEV _001	$-112.0 \leq RTWP < -111.9$	dBm
RTWP_LEV _002	$-111.9 \leq RTWP < -111.8$	dBm
. . .	. . .	. . .
RTWP_LEV _619	$-50.2 \leq RTWP < -50.1$	dBm
RTWP_LEV _620	$-50.1 \leq RTWP < -50.0$	dBm
RTWP_LEV _621	$-50.0 \leq RTWP$	dBm

All NBAP dedicated measurements report the quality of data transmission using dedicated channels on the uplink frequency band. Downlink quality measurements are reported by UEs using RRC measurement reports.

For NBAP dedicated measurements the same reporting event definitions apply as for common measurements (see Table 2.7). Some exceptions, e.g. for best cell portions reporting, are described in 3GPP 25.433. Apart from reporting events there is another important setting in the dedicated measurement object type that specifies if the measurement result has been measured on a single radio link (dedicated measurement object type = 'RL'), on all radio links for the same UE within a Node B (dedicated measurement object type = 'All RL') or for one or more radio link sets of a single UE (dedicated measurement object type = 'RLS'). It is important to know the settings before the evaluation of measurement results. Also the choice of a possible aggregation level depends on these settings.

Additionally, 3GPP 25.215 defines that for some measurement reports (e.g. SIR) not just a cell, but a cell portion can be indicated. The cell portion is a geographical part of a cell for which a cell portion ID (NBAP parameter) is assigned. These cell portions do not necessarily need to be identical with the beams of beam-forming antennas (if such antennas are installed in UTRAN), but the RNC may associate a cell portion with dedicated physical channels if the UE is located in a known cell portion.

The following NBAP dedicated measurements apply to FDD mode and will be described later in this section:

- signal-to-interference-ratio (SIR)
- signal-to-interference-ratio error (SIR error)
- transmitted code power
- round trip time (RTT)

All these measurements are basically initialised in the same way as shown in the signalling call flow of Figure 2.12. As a prerequisite an NBAP radio link setup procedure must have been executed successfully.

2.2.3.1 Signal-to-Interference Ratio (SIR)

SIR is the measured uplink quality of a single call. It is the ratio between the measured received signal code power (RSCP) of a single UE's uplink DPCCH's signal and the interference signal code power (ISCP) multiplied with the spreading factor of the DPCCH, which has a constant value of 256.

$$\text{SIR} = \frac{RSCP \ of \ DPCCH}{ISCP} \times 256 \tag{2.2}$$

ISCP is that part of the RTWP that is caused by uplink transmissions of other UEs using the same cell. ISCP is measured based on sophisticated proprietary algorithms implemented in Node B software. SIR has an impact on closed loop power control as shown in Figure 2.7 and if such (once again proprietary) algorithms are implemented in the SRNC software, SIR measurement reports can trigger handover decisions of the SRNC due to changes of uplink quality. However, most handover decisions are based on RRC

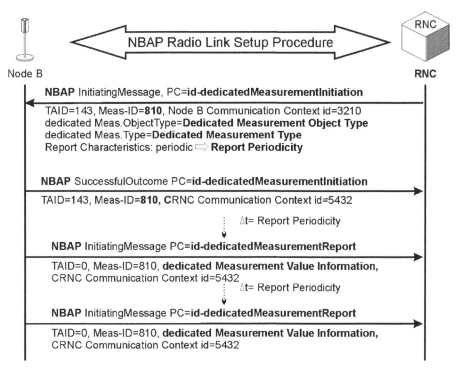

Figure 2.12 NBAP dedicated measurement initialisation, periodical reporting and termination

measurement reports sent by the UE, but RRC measurements only indicate the downlink quality of radio transmission.

SIR is reported in 64 bins. Each bin represents a range of 0.5 dB as shown in the Table 2.9.

2.2.3.2 Signal-to-Interference Ratio Error (SIR Error)

SIR error is the difference between the measured SIR and the average SIR target for a given sampling period.

$$\mathrm{SIR}_{error} = \mathrm{SIR} - \mathrm{SIR}_{target\ average} \tag{2.3}$$

Table 2.9 Bin mapping table for SIR (from 3GPP 25.133)

Reported value	Measured quantity value	Unit
UTRAN_SIR_00	SIR < −11.0	dB
UTRAN_SIR_01	−11.0 ≤ SIR < −10.5	dB
UTRAN_SIR_02	−10.5 ≤ SIR < −10.0	dB
. . .	. . .	. . .
UTRAN_SIR_61	19.0 ≤ SIR < 19.5	dB
UTRAN_SIR_62	19.5 ≤ SIR < 20.0	dB
UTRAN_SIR_63	20.0 ≤ SIR	dB

It is necessary to compute the $SIR_{target\_average}$ because new SIR targets might be assigned by the RNC during the sampling period. In compressed mode the $SIR_{target\_average}$ is not calculated over the transmission gap. The averaging of SIR_{target} is made in a linear scale and $SIR_{target\_average}$ is given in dB.

Message example 2.4 shows a measurement initialisation for an event-triggered SIR error reporting after event F. Periodical reporting is activated if SIR error exceeds (negative) measurement threshold 1 for more than 60 milliseconds (step size of hysteresis time $= 10\,\mathrm{ms}$) stopped according to the definition of measurement threshold 2 and hysteresis time.

SIR error is reported and initiated using bins as well. For mapping information of integer values to dB see Table 2.10.

Message example 2.4 NBAP dedicated measurement initialisation for SIR error following event F

ID Name	Comment or Value	
TS 25.433 V3.7.0 (2001-09) (NBAP)	initiatingMessage (= initiatingMessage)	
nbapPDU		
1 initiatingMessage		
1.1 procedureID		
1.1.1 procedureCode	id-dedicatedMeasurementInitiation	
1.5.1.1.1 id	id-NodeB-CommunicationContextID	
1.5.1.1.2 criticality	reject	
1.5.1.1.3 value	4	
1.5.1.2 sequence		
1.5.1.2.1 id	id-MeasurementID	
1.5.1.2.2 criticality	reject	
1.5.1.2.3 value	1	
1.5.1.3 sequence		
1.5.1.3.1 id	id-DedicatedMeasurementObjectType-DM-Rqst	
1.5.1.3.2 criticality	reject	
1.5.1.3.3 value		
1.5.1.3.3.1 **all-RL**	0	
1.5.1.4 sequence		
1.5.1.4.1 id	**id-DedicatedMeasurementType**	
1.5.1.4.2 criticality	reject	
1.5.1.4.3 value	**sir-error**	
1.5.1.6.1 id	id-ReportCharacteristics	
1.5.1.6.2 criticality	reject	
1.5.1.6.3 value		
1.5.1.6.3.1 **event-f**		
1.5.1.6.3.1.1 **measurementThreshold1**		
1.5.1.6.3.1.1.1 **sir-error**	59	
1.5.1.6.3.1.2 **measurementThreshold2**		
1.5.1.6.3.1.2.1 **sir-error**	58	
1.5.1.6.3.1.3 **measurementHysteresisTime**		
1.5.1.6.3.1.3.1 **msec**	6	

Table 2.10 Bin mapping table for SIR error (from 3GPP 25.133)

Reported value	Measured quantity value	Unit
UTRAN_SIR_ERROR_000	$SIR_{error} < -31.0$	dB
UTRAN_SIR_ERROR_001	$-31.0 \leq SIR_{error} < -30.5$	dB
UTRAN_SIR_ERROR_002	$-30.5 \leq SIR_{error} < -30.0$	dB
. . .	. . .	. . .
UTRAN_SIR_ERROR_062	$-0.5 \leq SIR_{error} < 0.0$	dB
UTRAN_SIR_ERROR_063	$0.0 \leq SIR_{error} < 0.5$	dB
. . .	. . .	. . .
UTRAN_SIR_ERROR_123	$30.0 \leq SIR_{error} < 30.5$	dB
UTRAN_SIR_ERROR_124	$30.5 \leq SIR_{error} < 31.0$	dB
UTRAN_SIR_ERROR_125	$31.0 \leq SIR_{error}$	dB

2.2.3.3 Uplink SIR Target

The uplink SIR target completes the SIR family quality parameters although it does not report radio link quality. It is sent as a command from the RNC to serving Node Bs using both NBAP and Iub FP signalling messages.

Note: although SIR target is sent in the downlink direction from the CRNC to Node B it is called the uplink SIR target because it defines the expected quality on uplink dedicated physical channels. Sometimes this is misunderstood and the term 'downlink SIR target"' is used, but there is no SIR measurement and also no SIR target defined on downlink dedicated physical channels in FDD mode.

The uplink SIR target is used to order Node B and its cells involved in radio transmission of uplink dedicated physical channels to ensure that the receiving signal quality on the uplink remains constant. Node B has an influence on the uplink transmission power of the UE by sending transmit power commands (TPC) to the mobile. According to those TPCs the UE will increase or decrease its uplink transmission power. The level of UE transmission power is reported to the SRNC using RRC measurement reports and appropriate event triggers are defined in RRC event group 6 (see section 2.2.4).

The uplink SIR target is the same for all Node Bs/cells involved in an active set, but transmit power commands sent by these Node Bs/cells may differ depending on the current position of a UE. It is the job of the UE to combine these multiple TPCs into a single command that determines if UE Tx power is increased or decreased.

Due to this, the uplink SIR target can be aggregated very well on call or UE level and changes in the uplink SIR target can be displayed in a time-based analysis diagram. Due to the fact that changed the uplink SIR target results in changed uplink transmission power of a UE, a number of uplink radio link quality parameters can be correlated to the uplink SIR target to prove if call quality has become better or worse by changing the target value. Such uplink radio quality parameters are especially the uplink bit error ratio encoded as a quality estimate and the uplink block error ratio (UL BLER) as described in section 2.1.1. In some cases instead of UL BLER the number of UL block CRC errors is shown in correlated call quality analysis, which also seems to be a useful approach.

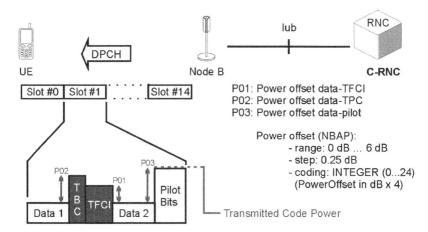

Figure 2.13   Measurement of transmitted code power

2.2.3.4 Transmitted Code Power

Transmitted code power is the used transmission power on a downlink DPCH for a single UE.

Transmitted code power reports may be used to trigger initialisation of UE measurements (RRC Measurement Control message) on inter-frequency or inter-RAT neighbour cells.

The measurement is done on the downlink DPCH pilot bits. During the set up of a radio link the RNC indicates three power offsets, PO1, PO2 and PO3, to Node B. The parameters can be found in the NBAP Initiating Message of the radio link setup procedure. For transmitted code power measurement PO3 is important. It indicates the difference of transmission power between the data part (coded DCH bits) relative to the power of the DPCH pilot bit field. PO3 can be set in a range between 0 dB and 6 dB in 0.25 dB steps. The transmitted code power is the power of the pilot bits. Using the power offset it is possible to determine the power of each of the various bit fields in a DPCH slot. See Figure 2.13.

Transmitted code power is reported in bins. Each bin covers a range of 0.5 dBm. Mapping between integer bin values and dBm ranges is shown in Table 2.11.

Table 2.11   Bin mapping table for transmitted code power (from 3GPP 25.133)

Reported value	Measured quantity value	Unit
UTRAN_CODE_POWER _010	$-10.0 \leq$ Transmitted code power < -9.5	dBm
UTRAN_CODE_POWER _011	$-9.5 \leq$ Transmitted code power < -9.0	dBm
UTRAN_CODE_POWER _012	$-9.0 \leq$ Transmitted code power < -8.5	dBm
$\cdots$	$\cdots$	$\cdots$
UTRAN_CODE_POWER _120	$45.0 \leq$ Transmitted code power < 45.5	dBm
UTRAN_CODE_POWER _121	$45.5 \leq$ Transmitted code power < 46.0	dBm
UTRAN_CODE_POWER _122	$46.0 \leq$ Transmitted code power < 46.5	dBm

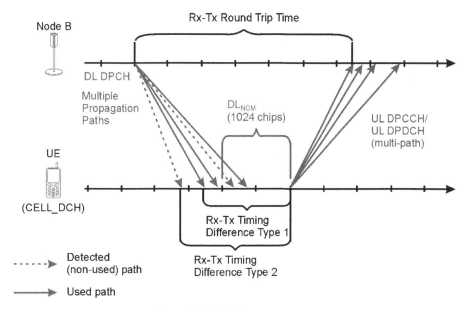

Figure 2.14   RTT measurements

2.2.3.5 Round Trip Time (RTT)

Basically there are RTT measurements defined for Node B and UE. Reports from the UE are monitored on the RRC protocol layer while Node B uses NBAP dedicated measurement reports.

Round trip time on the Uu interface as illustrated in Figure 2.14 is defined as the time between sending a downlink frame and receiving the appropriate uplink frame on the first detected path in the case of multi-path receiving.

Multi-path means that Node B (and UE as well) are able to not only receive the signal from direct radio connection, but also signals reflected by e.g. buildings. These reflected signals will usually arrive a little bit later than the direct radio connection signal at the antenna, but the rake receiver and matched filter in the UE and Node B are able to filter and combine the same signal received via multiple paths and combine single-path signal components into one strong output signal. This helps to minimise transmission power and hence to minimise the total interference in the cell and to maximise cell capacity.

RTT is reported in bins, the true unit is the chip. Time to transmit a chip is constant, because in UTRAN FDD constant chip rate is 3.84 Mcps (megachips per second). Following this a single chip represents a time interval of approximately 0.2604167 µs. See Table 2.12.

2.2.4 RRC MEASUREMENTS AND UE MEASUREMENT ABILITIES

In general all measurements regarding the quality of downlink transmission via the radio interface are executed by the UE and reported using RRC measurement report messages. However, a UE does not have a choice what needs to be measured and what does not. Each

Table 2.12 Bin mapping table for round trip time (from 3GPP 25.133)

Reported value	Measured quantity value	Unit
RT_TIME_0000	Round trip time < 876.0000	chip
RT_TIME_0001	876.0000 ≤ Round trip time < 876.0625	chip
RT_TIME_0002	876.0625 ≤ Round trip time < 876.1250	chip
RT_TIME_0003	876.1250 ≤ Round trip time < 876.1875	chip
. . .	. . .	. . .
RT_TIME_32764	2922.6875 ≤ Round trip time < 2923.7500	chip
RT_TIME_32765	2923.7500 ≤ Round trip time < 2923.8125	chip
RT_TIME_32766	2923.8125 ≤ Round trip time < 2923.8750	chip
RT_TIME_32767	2923.8750 ≤ Round trip time	chip

UE is ordered by the SRNC which parameters are to be measured in which cell and if and how measurement reports need to be sent.

Before the UE requests the set up of a RRC connection it receives measurement requests distributed via the cell broadcast channel. In system information block 11 (SIB 11) of the broadcast control channel (BCCH) we find initial cell selection and cell reselection information. Based on this information the UE decides which of the cells that can be measured at a certain geographical location is currently the best cell. The UE will then start to send some preambles using the physical random access channel (PRACH) to this best cell and if one of these preambles is acknowledged in the acquisition indication channel (AICH) of the same cell, the UE will send an RRC Connection Request message using the random access channel (RACH) of this cell. The RACH is the transport channel mapped onto the physical channel PRACH.

If the RRC connection between UE and SRNC is established as requested the UE will most likely receive new measurement instructions via a DCCH.

A DCCH is also called a signalling radio bearer (SRB) using the terminology of the UMTS quality of service concept.

Measurement instructions for UEs can be grouped in seven main categories:

1. *Intra-frequency measurements* related to cells working on the same frequency as currently used cell(s). Measurement reports of this category are used to trigger soft and softer handover radio link additions and radio link deletions. They may also trigger serving HS-DSCH cell change for HSDPA-capable cells and inter-frequency hard handovers.
2. *Inter-frequency measurements* are related to cells belonging to the same UTRAN/operator network, but working on a different frequency than currently used cell(s). Measurement reports of this category are used to trigger measurements in compressed mode as well as inter-frequency hard handover procedures.
3. *Inter-RAT measurements* are related to cells belonging to the GSM/GERAN or CDMA2000 portion of the same or a different network operator's network. Measurement reports belonging to this category are used to trigger handover to GSM/GERAN and/or CDMA2000. Both GSM/GERAN and CDMA2000 are called different radio access technology (RAT).
4. *Traffic volume measurements* deal with measurement of RLC buffer size for uplink IP payload   transmissions.   These   measurements   are   used   to   trigger   radio   bearer

reconfigurations. In special cases, if the transport channel for the PS radio bearer is changed, those procedures are called channel-type switching.

5. *Measurement of DL transport channel quality.* This is based on counting downlink CRC errors for transport blocks sent on defined downlink transport channels (excluding FACH). These measurements have been already discussed in section 2.1.2.

6. *UE internal measurements.* Mostly these measurements correspond to NBAP common measurements. UE internal measurements category provides reports of downlink cell quality and uplink transmission power.

7. Category 7 deals with *UE positioning reporting* and does not need to be discussed in detail.

As in NBAP measurement reports can be sent periodically or an event can be triggered. To minimise load on the SRNC event-triggered reporting is preferred and 3GPP recommends using periodical reporting only in special situations, for instance if the SRNC is unable to start a handover procedure after receiving an appropriate event-triggered measurement report.

Indeed, reporting events are the most sticky portions of RRC measurement report messages. Because of this call traces are often analysed by looking at reported events and for instance event 1A became a synonym for soft handover radio link addition. Actually, this kind of analysis is not careful enough, because the reported event is not a handover itself. It is a measurement, and based on this measurement and ability of resources the SRNC decides if a handover will subsequently be performed or not. This fact becomes immediately evident if periodical reporting is activated, because event reports are no longer sent, and the SRNC will continue to perform the same handover procedures as before – now based on reported measurement values. The second reason why the synonym should not be used is because the type of handover cannot be bound to a certain measurement event. In most observed cases event 1A triggers radio link addition for soft or softer handover, except in those cases when the new cell cannot be added to the existing radio link set, because this cell is controlled by another RNC and there is no Iur interface between the SRNC and the second one. Then an intra-frequency hard handover is performed, which is completely different from a radio link addition.

Basically event 1A says that 'a Primary CPICH enters the reporting range' (3GPP 25.331). This can be interpreted as 'another cell working on the same frequency as currently used becomes good enough so that a handover can be performed'.

To base a handover analysis only on measurement events is not a good idea, but it is useful to analyse measurement events related to signalling messages or message sequences to find out if e.g. an RRC physical channel reconfiguration procedure is used to perform hard handover or another type of reconfiguration.

Beside the sticky event reports there is a lot of other valuable measurement data found in RRC measurement report messages. In message example 2.5 the event report sequence is shown in light grey to switch focus onto reported measurement results.

Here it can be seen that the measurement report contains cell measured result sequences. For a number of neighbour cells the system frame number (SFN) timing differences are reported as specified in 3GPP 25.215. Then there is the primary scrambling code, the identifier of a UTRAN cell on the radio interface, and bin-encoded measurement values for chip energy over noise (Ec/N0) and received signal code power (RSCP) of the primary

Message example 2.5 RRC Intra-frequency measurement report

BITMASK	ID Name	Comment or Value
TS 25.331 DCCH-UL (2002-03) (RRC_DCCH_UL) measurementReport (= measurementReport)		
uL-DCCH-Message		
2.1 measurementReport		
b4 2.1.1 measurementIdentity	1	
2.1.2 measuredResults		
2.1.2.1 intraFreqMeasuredResultsList		
2.1.2.1.1 cellMeasuredResults		
2.1.2.1.1.1 cellSynchronisationInfo		
2.1.2.1.1.1.1 modeSpecificInfo		
2.1.2.1.1.1.1.1 fdd		
2.1.2.1.1.1.1.1.1 countC-SFN-Frame-difference		
0000 —— 2.1.2.1.1.1.1.1.1.1 countC-SFN-High	0	
b8 2.1.2.1.1.1.1.1.1.2 off	168	
b16* 2.1.2.1.1.1.1.1.1.2 tm	2092	
2.1.2.1.1.2 modeSpecificInfo		
2.1.2.1.1.2.1 fdd		
2.1.2.1.1.2.1.1 primaryCPICH-Info		
b9 2.1.2.1.1.2.1.1.1 primaryScramblingCode	20	
-011000- 2.1.2.1.1.2.1.1.2 cpich-Ec-N0	24	
b7 2.1.2.1.1.2.1.1.3 cpich-RSCP	11	
2.1.2.1.2 cellMeasuredResults		
2.1.2.1.2.1 cellSynchronisationInfo		
2.1.2.1.2.1.1 modeSpecificInfo		
2.1.2.1.2.1.1.1 fdd		
2.1.2.1.2.1.1.1.1 countC-SFN-Frame-difference		
—0000- 2.1.2.1.2.1.1.1.1.1 countC-SFN-High	0	
b8 2.1.2.1.2.1.1.1.1.2 off	183	
b16* 2.1.2.1.2.1.1.1.2 tm	20523	
2.1.2.1.2.2 modeSpecificInfo		
2.1.2.1.2.2.1 fdd		
2.1.2.1.2.2.1.1 primaryCPICH-Info		
b9 2.1.2.1.2.2.1.1.1 primaryScramblingCode	386	
b6 2.1.2.1.2.2.1.1.2 cpich-Ec-N0	24	
b7 2.1.2.1.2.2.1.1.3 cpich-RSCP	11	
2.1.3 eventResults		
2.1.3.1 intraFreqEventResults		
0010—— 2.1.3.1.1 eventID	e1c	
2.1.3.1.2 cellMeasurementEventResults		
2.1.3.1.2.1 fdd		
2.1.3.1.2.1.1 primaryCPICH-Info		
b9 2.1.3.1.2.1.1.1 primaryScramblingCode	361	
2.1.3.1.2.1.2 primaryCPICH-Info		
b9 2.1.3.1.2.1.2.1 primaryScramblingCode	280	

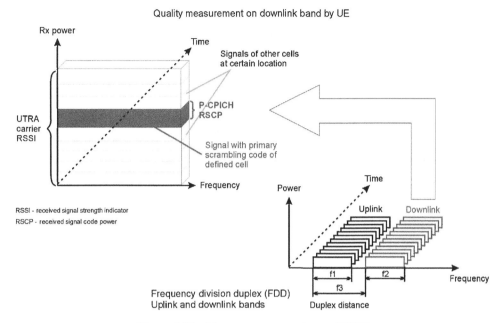

Figure 2.15 Measurements on primary CPICH

common pilot channel (P-CPICH) for each neighbour cell. Another value that could be reported is the UTRA carrier received signal strength indicator (UTRA RSSI).

These three measurements are related to each other as desribed in Equation (2.4) and Figure 2.15:

$$Ec/N0 = \frac{P\text{-CPICH RSCP}}{\text{UTRA Carrier RSSI}} \quad (2.4)$$

UTRA RSSI is so to speak the downlink equivalent to received total wideband power measured on the uplink and P-CPICH Ec/N0 could be seen as the downlink equivalent to the uplink SIR. The UE also computes a downlink SIR for the closed loop power control of the DPCH. However, details of this implementation are not specified by 3GPP for FDD mode and hence follow proprietary standards. The downlink SIR of the DPCH is not included in any measurement reports sent by the UE to the network.

The UE reports measurement results for the P-CPICH quality of all cells that can be received at a certain geographical location. A detailed analysis of this measurement inform-ation allows us not only to track location and changing radio environment conditions for a single UE, but also allows a cell-related neighbour analysis to find out which cells overlap/ interfere with each other and if there are any differences between planned and measured interference. Such analysis is mostly done using a so-called cell matrix. Different versions of such matrices, filter options and limitations will be discussed in Section 2.14.

Reported cells may belong to an active set of the UE or – if they have been members of a cell neighbour list sent to the UE, but currently not in an active set – they belong to the

Table 2.13 Bin mapping table for CPICH RSCP (from 3GPP 25.133)

Reported value	Measured quantity value	Unit
CPICH_RSCP_LEV _−05	CPICH RSCP < −120	dBm
CPICH_RSCP_LEV _−04	−120 ≤ CPICH RSCP < −119	dBm
CPICH_RSCP_LEV _−03	−119 ≤ CPICH RSCP < −118	dBm
. . .	. . .	. . .
CPICH_RSCP_LEV _89	−27 ≤ CPICH RSCP < −26	dBm
CPICH_RSCP_LEV _90	−26 ≤ CPICH RSCP < −25	dBm
CPICH_RSCP_LEV _91	−25 ≤ CPICH RSCP	dBm

monitored set. Finally, cells that are measured and reported without being members of the cell neighbour list belong to the detected set.

If reported cells can be identified as members of a detected set a discrepancy between radio network planning and reality is always indicated.

Due to the fact that the active set and cell neighbour lists constantly change because of the mobility of the UE a highly sophisticated software application is necessary to enable this analysis.

Another problem is that only 512 primary scrambling codes are available for a whole network with thousands of cells. Hence, the same primary scrambling code is used for many cells and the topology module needs to ensure that each cell can be identified uniquely. Otherwise there is a high risk that measurement results belonging to different cells that have the same primary scrambling code are mixed. If this happens all cell matrices, tables and diagrams derived from measurement values of RRC measurement reports become useless.

Once again values of UTRA carrier RSSI, RSCP and Ec/N0 are encoded in bins as described in Tables 2.13, 2.14 and 2.15.

It is an excellent approach to present values of these three radio quality parameters in cell matrices (see section 2.14). However, Ec/N0 values monitored in RRC Connection Request and RRC Cell Update messages will not be considered for cell matrix set up. The reason is that in these messages one sees only values of the best cells reported while other neighbour cells that are monitored by the UE will not be included. Nevertheless, the analysis of Ec/N0 measured for the best cell selected by the UE to start up a network connection allows statements if cells (re-)selected by UEs for initial access to the UTRAN offered a sufficient

Table 2.14 Bin mapping table for CPICH Ec/N0 (from 3GPP 25.133)

Reported value	Measured quantity value	Unit
CPICH_Ec/No _00	CPICH Ec/Io < −24	dB
CPICH_Ec/No _01	−24 ≤ CPICH Ec/Io < −23.5	dB
CPICH_Ec/No _02	−23.5 ≤ CPICH Ec/Io < −23	dB
. . .	. . .	. . .
CPICH_Ec/No _47	−1 ≤ CPICH Ec/Io < −0.5	dB
CPICH_Ec/No _48	−0.5 ≤ CPICH Ec/Io < 0	dB
CPICH_Ec/No _49	0 ≤ CPICH Ec/Io	dB

Table 2.15 Bin mapping table for UTRA carrier RSSI (from 3GPP 25.133)

Reported value	Measured quantity value	Unit
UTRA_carrier_RSSI_LEV _00	UTRA carrier RSSI < −100	dBm
UTRA_carrier_RSSI_LEV _01	−100 ≤ UTRA carrier RSSI < −99	dBm
UTRA_carrier_RSSI_LEV _02	−99 ≤ UTRA carrier RSSI < −98	dBm
. . .	. . .	. . .
UTRA_carrier_RSSI_LEV _74	−27 ≤ UTRA carrier RSSI < −26	dBm
UTRA_carrier_RSSI_LEV _75	−26 ≤ UTRA carrier RSSI < −25	dBm
UTRA_carrier_RSSI_LEV _76	−25 ≤ UTRA carrier RSSI	dBm

downlink quality. The best way of a graphical analysis is provided by a histogram that shows distribution of values. Due to the fact that different thresholds are defined for interRAT cell reselection compared to intial UTRAN cell selection (all other establishment causes like 'registration', 'originating conversational call' etc.) a separate analysis per establishment cause is useful. In addition it can be also of some interest to find out which Ec/N0 value in average was necessary for a certain type of UE to get successful access to network using RACH. For this reason it is necessary to correlated Ec/N0 values extracted from RRC Connection Request and Cell Update with IMEI of UEs (as far as IMEI is monitored in call flow). Leading six numbers of IMEI allow to determine the type of UE. For instance Nokia 6680 mobile phones can be identified by leading IMEI digits 355661 and 355664. Comparing Ec/N0 values reported on RACH it can be evaluated which UE types in average need a better quality of the cell signal to get successful access to the network. Since especially RRC connection setup failure rate in today's network is quite high (approx. 10% in peak times of traffic is seen as a common value) it is assumed that UEs reporting a better average Ec/N0 on RACH are the ones that more often fail to establish the RRC connection successfully. However, due to the fact that RRC connection setup failure events can never be correlated to an IMEI a final prove of this assumption using performance measurement software cannot be done. Last but not least it must also be kept in mind that different Ec/N0 thresholds defined for individual establishment/cell update causes will correspond to individual RRC Connection Setup/RRC Cell Update Success and Failure Rates per cause value. Typically an Ec/N0 value of −14 dB is seen as sufficient for inter-RAT cell reselection while for start of initial registration procedures −10 dB are required.

It is further interesting to have a closer look at some options defined for downlink radio quality parameter measurements. It is possible, for example, that within a cell there are several common pilot channels, one primary CPICH and up to 15 secondary CPICHs. Secondary CPICHs are used to define sub-cells within the primary cell using beam-forming antennas. If a UE is located in a sub-cell it is possible that appropriate measurement results of RSCP and Ec/N0 are reported based on the measurement of the secondary CPICH belonging to this sub-cell. Whether a secondary CPICH is used depends on two facts. First, using a special information element in the RRC protocol indicates that the 'Primary CPICH shall not be used for channel estimation' (3GPP 25.331). Secondly, information included in RRC messages used for RRC connection establishment, radio bearer set up or physical/transport channel/radio bearer reconfiguration needs to order the UE to use a sub-cell identified by the appropriate secondary CPICH. From the measurement value in the RRC

Message example 2.6 Compressed mode activation sequence in NBAP

ID Name	Comment or Value	
TS 25.433 V3.12.0 (NBAP) initiatingMessage (= initiatingMessage)		
nbapPDU		
1 initiatingMessage		
1.1 procedureID		
1.1.1 procedureCode	id-synchronisedRadioLinkRon-figurationPreparation	
1.5.1.2.3.1.3.2.1.3 **transmissionGapPatternSequenceCod**··	**code-change**	
1.5.1.3 sequence		
1.5.1.3.1 id	id-Transmission-Gap-Patte...	
1.5.1.3.2 criticality	reject	
1.5.1.3.3 value		
1.5.1.3.3.1 sequence		
1.5.1.3.3.1.1 tGPSID	4	
1.5.1.3.3.1.2 tGSN	8	
1.5.1.3.3.1.3 tGL1	14	
1.5.1.3.3.1.4 tGD	0	
1.5.1.3.3.1.5 tGPL1	4	
1.5.1.3.3.1.6 tGPL2	4	
1.5.1.3.3.1.7 **uL-DL-mode**	**both-ul-and-dl**	
1.5.1.3.3.1.8 **downlink-Compressed-Mode-Method**	**sFdiv2**	
1.5.1.3.3.1.9 **uplink-Compressed-Mode-Method**	**sFdiv2**	

measurement report it cannot be detected if the measurement has been done on the primary or secondary CPICH.

Reported values are also not strictly bound to intra-frequency measurements. It is also possible that the same measurement values are reported for a UTRAN cell working on a different frequency than the currently used one. The prerequisite for this kind of measurement is the so-called compressed mode.

In compressed mode radio transmission on the currently used radio link is frequently interrupted. Extremely short interruption periods are called gaps and are used to gather measurement results of neighbour cells working on different UTRAN frequencies or different RATs.

Whether the compressed mode for a defined radio link is activated can be detected very easily by looking at NBAP signalling, because there are special message sequences defined for compressed mode activation. However, the UE also needs to be informed about compressed mode activation using RRC signalling – otherwise it would probably be surprised that the downlink data stream on the radio link is periodically interrupted and it would not deliver any inter-frequency or inter-RAT measurement results.

Compressed mode is of course not a standard mode for radio transmission and as a rule it needs a special measurement event to trigger the activation of compressed mode measurements. If event-triggered measurement reporting is activated this event is usually described by event-ID 2D: 'The estimated quality of the currently used frequency is below a certain threshold'.

Message example 2.7 Inter-frequency reporting set up

ID Name	Comment or Value
TS 25.331 DCCH-DL - V3.13.0 (RRC_DCCH_DL) measurementControl (= measurementControl)	
dL-DCCH-Message	
2 message	
2.1 measurementControl	
2.1.1.1.3.1 setup	
2.1.1.1.3.1.1 interFrequencyMeasurement	
2.1.1.1.3.1.1.1 interFreqCellInfoList	
2.1.1.1.3.1.1.2 interFreqMeasQuantity	
2.1.1.1.3.1.1.2.1 reportingCriteria	
2.1.1.1.3.1.1.2.1.1 interFreqReportingCriteria	
2.1.1.1.3.1.1.2.1.1.2.1.1 **freqQualityEstimateQuantity..**	**cpich-RSCP**
2.1.1.1.3.1.1.3 interFreqReportingQuantity	
2.1.1.1.3.1.1.3.1 **utra-Carrier-RSSI**	**false**
2.1.1.1.3.1.1.3.2 **frequencyQualityEstimate**	**true**
2.1.1.1.3.1.1.3.3 nonFreqRelatedQuantities	
2.1.1.1.3.1.1.3.3.1 dummy	noReport
2.1.1.1.3.1.1.3.3.2 cellIdentity-reportingIndicator	false
2.1.1.1.3.1.1.3.3.3 cellSynchronisationInfoReportingI..	false
2.1.1.1.3.1.1.3.3.4 modeSpecificInfo	
2.1.1.1.3.1.1.3.3.4.1 fdd	
2.1.1.1.3.1.1.3.3.4.1.1 **cpich-Ec-N0-reportingIndicator**	**true**
2.1.1.1.3.1.1.3.3.4.1.2 **cpich-RSCP-reportingIndicator**	**true**
2.1.1.1.3.1.1.3.3.4.1.3 **pathloss-reportingIndicator**	**true**

Note: if there are problems on the UE or Node B side to activate compressed mode it often results in radio link failures or unrecoverable CRC errors reported. Hence, it is useful to define subsets of failure counters indicating if compressed mode has been activated before reported errors or not. For failure event definition see section 2.13.

In the case of inter-frequency measurements, measurement parameters are identical to intra-frequency measurement results; the quality estimation of a cell working on a different RAT like GSM is totally different. For this reason the measurement of GSM carrier RSSI is defined. This measurement allows the evaluation of strength of the neighbour GSM cell. The identity of a GSM cell is provided by the base station identification code (BSIC), which should be ideally be identified by the UE during measurement in compressed mode.

BSIC consists of the network colour code (NCC) and the base station colour code (BCC). Together with the absolute radio frequency channel number (ARFCN), which allows the UE to calculate the frequency on which a defined GSM sends its training sequence and the broadcast channel, they build a unique triplet. BCC and NCC allow a unique identification of a GSM cell, because cells that have the same ARFCN may belong to different operators' networks. However, in this case BSIC is different.

If BSIC identification is not possible the cell is either approved to be a possible handover target or not. It would be an interesting investigation to find out if the risk that handover to GSM fails is higher if the identity of the target cell is not known.

Message example 2.8 Inter-RAT measurement set up

ID Name	Comment or Value	
TS 25.331 DCCH-DL (2002-03) (RRC_DCCH_DL) measurementControl (= measurementControl)		
2 message		
2.1 measurementControl		
2.1.1.1.3 measurementCommand		
2.1.1.1.3.1 setup		
2.1.1.1.3.1.1 interRATMeasurement		
2.1.1.1.3.1.1.1 interRATCellInfoList		
2.1.1.1.3.1.1.1.2.1 newInterRATCell		
2.1.1.1.3.1.1.1.2.1.1 interRATCellID	0	
2.1.1.1.3.1.1.1.2.1.2 technologySpecificInfo		
2.1.1.1.3.1.1.1.2.1.2.1 gsm		
2.1.1.1.3.1.1.1.2.1.2.1.1 interRATCellIndividualOffset	0	
2.1.1.1.3.1.1.1.2.1.2.1.2 bsic		
2.1.1.1.3.1.1.1.2.1.2.1.2.1 ncc	1	
2.1.1.1.3.1.1.1.2.1.2.1.2.2 bcc	5	
2.1.1.1.3.1.1.1.2.1.2.1.3 frequency-band	dcs1800BandUsed	
2.1.1.1.3.1.1.1.2.1.2.1.4 bcch-ARFCN	79	
2.1.1.1.3.1.1.1.2.2 newInterRATCell		
2.1.1.1.3.1.1.1.2.2.1 interRATCellID	1	
2.1.1.1.3.1.1.1.2.2.2 technologySpecificInfo		
2.1.1.1.3.1.1.1.2.2.2.1 gsm		
2.1.1.1.3.1.1.2 interRATMeasQuantity		
2.1.1.1.3.1.1.2.1 measQuantityUTRAN-QualityEstimate		
2.1.1.1.3.1.1.2.1.2.1.1 intraFreqMeasQuantity-FDD	cpich-RSCP	
2.1.1.1.3.1.1.2.2 ratSpecificInfo		
2.1.1.1.3.1.1.2.2.1 gsm		
2.1.1.1.3.1.1.2.2.1.1 measurementQuantity	gsm-CarrierRSSI	
2.1.1.1.3.1.1.2.2.1.2 filterCoefficient	fc0	
2.1.1.1.3.1.1.2.2.1.3 bsic-VerificationRequired	required	
2.1.1.1.3.1.1.3 interRATReportingQuantity		
2.1.1.1.3.1.1.3.1 utran-EstimatedQuality	false	
2.1.1.1.3.1.1.3.2 ratSpecificInfo		
2.1.1.1.3.1.1.3.2.1 gsm		
2.1.1.1.3.1.1.3.2.1.1 dummy	false	
2.1.1.1.3.1.1.3.2.1.2 observedTimeDifferenceGSM	false	
2.1.1.1.3.1.1.3.2.1.3 gsm-Carrier-RSSI	true	
2.1.1.1.3.1.1.4 reportCriteria		
2.1.1.1.3.1.1.4.1 interRATReportingCriteria		
2.1.1.1.3.1.1.4.1.1 interRATEventList		
2.1.1.1.3.1.1.4.1.1.1 interRATEvent		
2.1.1.1.3.1.1.4.1.1.1.1 event3a		

Message example 2.9 RRC measurement report with inter-RAT measurement results

ID Name	Comment or Value	
TS 25.331 DCCH-UL (2002-03) (RRC_DCCH_UL)	measurementReport (= measurementReport)	
uL-DCCH-Message		
2 message		
2.1 measurementReport		
2.1.1 measurementIdentity	3	
2.1.2 measuredResults		
2.1.2.1 interRATMeasuredResultsList		
2.1.2.1.1 interRATMeasuredResults		
2.1.2.1.1.1 gsm		
2.1.2.1.1.1.1 gSM-MeasuredResults		
2.1.2.1.1.1.1.1 gsm-CarrierRSSI	'100100'B	
2.1.2.1.1.1.1.2 bsicReported		
2.1.2.1.1.1.1.2.1 nonVerifiedBSIC	85	
2.1.2.1.1.1.2 gSM-MeasuredResults		
2.1.2.1.1.1.2.1 gsm-CarrierRSSI	'100001'B	
2.1.2.1.1.1.2.2 bsicReported		
2.1.2.1.1.1.2.2.1 nonVerifiedBSIC	75	
2.1.3 eventResults		
2.1.3.1 interRATEventResults		
2.1.3.1.1 eventID	e3a	
2.1.3.1.2 cellToReportList		
2.1.3.1.2.1 cellToReport		
2.1.3.1.2.1.1 bsicReported		
2.1.3.1.2.1.1.1 verifiedBSIC	2	

In RRC measurement reporting group number 4, which deals with traffic volume measurement, it is also possible to get reports about buffer volumes for uplink data transmission on the UE side. However, currently reports are mostly limited to threshold reporting using event-IDs.

Message example 2.10 shows the activation of traffic volume reporting for a radio bearer using either DCH #21, RACH or CPCH. The real payload buffer size is not reported, but event-IDs 4A (buffer above defined threshold) and 4B (buffer below defined threshold) are reported to indicate buffer usage and trigger the necessary reconfiguration of transport channels, i.e. to switch from using DCH to RACH/FACH and vice versa. While most RRC measurements are only reported when the UE is in CELL_DCH state traffic volume reporting is also enabled if the UE is in CELL_FACH state.

The same choice is possible between event-triggered reporting and periodic reporting in the case of DL quality measurements. DL quality reported by the UE is the DL BLER for each used downlink dedicated transport channel. Transport channels mapped onto secondary common control physical channels (S-CCPCH), which means e.g. FACH, are excluded from this kind of measurement. Message examples for RRC Measurement Control/RRC Measurement Report with DL transport channel BLER reporting have already been shown in Section 2.1.

Message example 2.10 Set up of traffic volume reporting

ID Name	Comment or Value
TS 29.331 DCCH-DL (2002-09) (RRC_DCCH_DL) measurementControl (= measurementControl)	
dL-DCCH-Message	
2 message	
2.1 measurementControl	
2.1.1.1.3.1 setup	
2.1.1.1.3.1.1 trafficVolumeMeasurement	
2.1.1.1.3.1.1.1 trafficVolumeMeasurementObjectList	
2.1.1.1.3.1.1.1.1 uL-TrCH-Identity	
2.1.1.1.3.1.1.1.1.1 **dch**	**21**
2.1.1.1.3.1.1.1.2 uL-TrCH-Identity	
2.1.1.1.3.1.1.1.2.1 **rachorcpch**	**0**
2.1.1.1.3.1.1.2 trafficVolumeMeasQuantity	
2.1.1.1.3.1.1.2.1 averageRLC-BufferPayload	7
2.1.1.1.3.1.1.3 trafficVolumeReportingQuantity	
2.1.1.1.3.1.1.3.1 **rlc-RB-BufferPayload**	**false**
2.1.1.1.3.1.1.3.2 **rlc-RB-BufferPayloadAverage**	**false**
2.1.1.1.3.1.1.3.3 **rlc-RB-BufferPayloadVariance**	**false**
2.1.1.1.3.1.1.4 measurementValidity	
2.1.1.1.3.1.1.4.1 **ue-State**	**all-States**
2.1.1.1.3.1.1.5 reportCriteria	
2.1.1.1.3.1.1.5.1 trafficVolumeReportingCriteria	
2.1.1.1.3.1.1.5.1.1 transChCriteriaList	
2.1.1.1.3.1.1.5.1.1.1 transChCriteria	
2.1.1.1.3.1.1.5.1.1.1.1 ul-transportChannelID	
2.1.1.1.3.1.1.5.1.1.1.1.1 **dch**	**21**
2.1.1.1.3.1.1.5.1.1.1.2 eventSpecificParameters	
2.1.1.1.3.1.1.5.1.1.1.2.1 trafficVolumeEventParam	
2.1.1.1.3.1.1.5.1.1.1.2.1.1 **eventID**	**e4b**
2.1.1.1.3.1.1.5.1.1.1.2.1.2 reportingThreshold	th8k
2.1.1.1.3.1.1.5.1.1.1.2.1.3 timeToTrigger	ttt1280
2.1.1.1.3.1.1.5.1.1.1.2.1.4 pendingTimeAfterTri..	ptat2
2.1.1.1.3.1.1.5.1.1.2 transChCriteria	
2.1.1.1.3.1.1.5.1.1.2.1 ul-transportChannelID	
2.1.1.1.3.1.1.5.1.1.2.1.1 rachorcpch	0
2.1.1.1.3.1.1.5.1.1.2.2 eventSpecificParameters	
2.1.1.1.3.1.1.5.1.1.2.2.1 trafficVolumeEventParam	
2.1.1.1.3.1.1.5.1.1.2.2.1.1 **eventID**	**e4a**
2.1.1.1.3.1.1.5.1.1.2.2.1.2 reportingThreshold	th1024
2.1.1.1.3.1.1.5.1.1.2.2.1.3 timeToTrigger	ttt20
2.1.1.1.3.1.1.5.1.1.2.2.1.4 pendingTimeAfterTri..	ptat2
2.1.1.1.3.1.1.5.1.1.2.2.1.5 tx-InterruptionAfte..	txiat2
2.1.1.1.4 measurementReportingMode	
2.1.1.1.4.1 measurementReportTransferMode	acknowledgedModeRLC
2.1.1.1.4.2 **periodicalOrEventTrigger**	**eventTrigger**

Message example 2.11 Measurement report for UE Tx power

ID Name	Comment or Value	
TS 29.331 DCCH-UL (2002-09) (RRC_DCCH_UL) measurementReport (= measurementReport)		
uL-DCCH-Message		
2 message		
2.1 measurementReport		
2.1.1 measurementIdentity	10	
2.1.2 measuredResults		
2.1.2.1 ue-InternalMeasuredResults		
2.1.2.1.1 modeSpecificInfo		
2.1.2.1.1.1 fdd		
2.1.2.1.1.1.1 **ue-TransmittedPowerFDD**	82	
2.1.3 eventResults		
2.1.3.1 ue-InternalEventResults		
2.1.3.1.1 **event 6b**	0	

Table 2.16 Bin mapping table for UE transmitted power measurement (from 3GPP 25.133)

Reported value	Measured quantity value (dBm)	Accuracy (dB) (note 1)	
UE_TX_POWER _104	33<= to <34	note 2	
UE_TX_POWER _103	32<= to <33	note 2	
UE_TX_POWER _102	31<= to <32	note 2	
. . . . . .			
UE_TX_POWER _096	25<= to <26	note 2	
UE_TX_POWER _095	24<= to <25	2.0	−2.0
UE_TX_POWER _094	23<= to <24	2.0	−2.0
UE_TX_POWER _093	22<= to <23	2.0	−2.0
UE_TX_POWER _092	21<= to <22	2.0	−2.0
UE_TX_POWER _091	20<= to < 21	2.5	−2.5
UE_TX_POWER _090	19<= to <20	3.0	−3.0
UE_TX_POWER _089	18<= to <19	3.5	−3.5
UE_TX_POWER _088	17<= to <18	4.0	−4.0
UE_TX_POWER _087	16<= to <17	4.0	−4.0
UE_TX_POWER _086	15<= to <16	4.0	−4.0
UE_TX_POWER _085	14<= to <15	4.0	−4.0
UE_TX_POWER _084	13<= to <14	4.0*	−4.0*
UE_TX_POWER _083	12<= to <13	4.0*	−4.0*
UE_TX_POWER _082	11<= to <12	4.0*	−4.0*
UE_TX_POWER _081	10<= to <11	note 2	
. . . . . .			
UE_TX_POWER _023	−48<= to < −47	note 2	
UE_TX_POWER _022	−49<= to < −48	note 2	
UE_TX_POWER _021	−50<= to < −49	note 2	

Note 1: the tolerance is specified for the maximum and minimum measured quantity value (dBm), i.e. MIN(Measured quantity value) + MIN(Accuracy) < = UE transmitted Power < Max (Measured quantity value) + MAX(Accuracy)

Note 2: no tolerance is specified.

Note*: applicable to power class 4

For deeper analysis it might be interesting to note that the setup message does not contain a transport channel ID. The allocation of quality measurement tasks to new transport channels added to active connection is done automatically. This means as long as the only connection between an UE and the network is a transport channel for stand-along signalling radio bearers there is also only one downlink BLER value reported that is related to this single DCH. After radio bearer set up the quality reporting for all payload DCHs that have been added to the connection is started immediately. Now each RRC Measurement Report contains as many DL BLER measuremet results as dedicated transport channels are active for the connection. Another limitation is that measurement values can only be reported for those DCHs that have a CRC check defined in their transport format settings.

The measurement set up may also order the UE to measure its transmitted power, which is the uplink equivalent to the transmitted code power of the cell reported using the NBAP common measurement report. Event-IDs 6A and 6B indicate that UE Tx power becomes more or less than a defined threshold. In RRC measurement report message example 2.11 threshold values as well as measurement results are encoded in bins as defined in 3GPP 25.133. This measurement report shows the UE Tx power is in range between 11 and 12 dB and is below a certain threshold defined for event-ID 6B.

As one can see in Table 2.16 a standard measurement tolerance is defined. These measurement tolerances (or as a confirmed sceptic would say: 'measurement failures') will be discussed in the following sections.

2.3 THROUGHPUT MEASUREMENTS

Throughput measurements can be done in many different ways on many different levels. This section will introduce the most common requirements.

The goal is to look at throughput measured on four different protocol levels and explain specific problems of correlation and aggregation. The four basic and most common throughput measurements for UTRAN analysis are:

- RLC throughput per cell. This is the throughput of all RLC transport blocks per cell, no matter if transport blocks carry user plane or control plane data or if they are retransmitted or not.
- Transport channel throughput per call is especially valuable for those dedicated transport channels that carry IP payload. To evaluate if measured throughput results are satisfying or not it is necessary to know the maximum possible throughput for a transport channel at a certain time in a call, because this maximum is a variable due to different reconfiguration strategies.
- User perceived throughput per call delivers results of filtered user plane traffic throughput without retransmissions, RLC/MAC header information and padding bits.
- Application throughput is measured on the application layer of the TCP/IP protocol stack and used to evaluate e.g. transmission speed for file transfer and video streaming applications.

Aggregating the last three throughput measurements at cell level is only possible using special implementations and filtering procedures that will be discussed in the following sections.

2.3.1 RLC THROUGHPUT

The first problem with measuring RLC throughput in a cell is related to software architecture. As discussed in chapter 1 all call-related data is filtered using the call trace application. However, not all RLC traffic in a cell is call related. Nor can all RLC traffic on radio interface be measured on the Iub interface. Hence, there will always be a difference between real RLC throughput on the Uu interface and RLC throughput measured on the Iub interface.

This statement addresses the difficulty of handling broadcasted information sent on the BCH and the paging channel (PCH). On the BCH system information is sent frequently in the downlink direction encoded in system information blocks (SIBs). These SIBs contain RRC signalling information and are constructed by the CRNC. However, since these SIBs are sent frequently on the radio interface they are stored in Node B according to cells. Due to this concept most SIBs on the Iub interface can only be monitored during Node B setup or reset procedure. In contrast to other RRC information this special kind of RRC signalling is not transmitted on separate Iub physical transport bearers (AAL2 SVC), it is sent across the Iub interface piggybacked on NBAP (!) messages. To learn more about the details of this transport procedure it is recommended to read the chapter about Node B setup scenario in Kreher and Ruedebusch (2005).

In addition to SIBs sent on the BCH, the PCH is also used to send system information to UEs camping on a cell. In this scenario not every paging message contains real paging information. Paging messages can also be used to distribute information about when and how to read data from the BCH. These paging messages cannot be filtered by a call trace application.

Note: many call trace applications currently used in monitoring equipment do not filter paging messages belonging to mobile terminated connections and show paging response as the first message of such connections instead.

All in all these known problems regarding the measurement of RLC throughput per cell (and some similar cell-related measurements) require an application that is able to filter RLC frames transported on all transport channels belonging to a single cell. Which channel belongs to which cell needs to be detected by a topology module. A software module that filters traffic from these channels is often called a cell trace application.

Channels that are involved in cell trace filtering are:

- random access channel (RACH)
- if available in cell: common packet channel (CPCH)
- variable number of forward access channels (FACH)
- paging channel (PCH)
- variable number of dedicated transport channels (DCH)
- if available in cell: high speed downlink shared channel(s) (HS-DSCH)

The bit rate on the HS-DSCH must be taken from NBAP common measurement reports and cannot be measured on the Iub interface. This is because retransmissions of hybrid automatic repeat request (HARQ) error correction require a fairly high percentage of HS-DSCH throughput, but HARQ only works on the radio interface between the UE and Node B.

The BCH is beyond the scope of cell trace application and therefore it must be carefully considered if it is really worthwhile spending a lot of effort on the development and testing of cell trace software compared to the value of better, but still not absolutely precise measurement results.

2.3.2 TRANSPORT CHANNEL THROUGHPUT

Transport channel throughput can also be seen as a kind of RLC throughput, but related to single calls/connections. It especially makes sense to measure it on dedicated transport channels. Throughput measurement on RACH and FACH is also possible, but does not increase in value as long as the transport channel usage ratio is not computed (see later) due to the mix of signalling and IP payload on RACH and FACH as discussed in the section about PS call scenarios in chapter 1 of this book.

Separate dedicated transport channels are used to transport SRBs and radio bearers (RBs) across the radio interface. Radio bearers are used for the transport of user plane data, e.g. speech and PS data.

To analyse the throughput of the DCHs carrying RRC signalling is not very significant, because the bandwidth of these DCHs is quite low (typically 1.7 or 3.4 kbps) and how much signalling data is transported using these channels depends on how many signalling messages need to be exchanged between the UE and the network. If the UE does not change its location and does not periodically send measurement reports there is not much signalling to be transmitted and hence the transport channel throughput to be measured is quite low.

Measurement on the DCHs used to transport AMR speech packets is also not very meaningful, because due to the nature of the AMR codec the provided bandwidth of the transport channels is either fully loaded, if there are voice packets to be transmitted, or the throughput is very low if voices are silent and only a few frames containing artificial noise are sent. Under these circumstances high or low throughput only indicates how many words have been spoken.

To put it in a nutshell, throughput measurement on signalling transport channels and on speech transport channels does not make sense, but a good approach is to measure the throughput for transport channels that carry IP payload. These measurements are really important, because the transmission speed is one of the most crucial performance parameters of customer-perceived quality of service, and if there are problems with transmission speeds these problems will most likely have their origin on the radio interface since this is the natural bottleneck for all kinds of data transmission using UMTS. However, to determine if the bottleneck has really been on the radio interface or somewhere else in UTRAN it is necessary to compare throughput measurement results with transport channel parameter settings.

Note: there is no value for any transport channel throughput measurement without knowing and comparing transport channel parameter settings with measurement results.

The theoretically possible data transmission rate or throughput on a UTRAN DCH is defined as follows:

$$\frac{Max\ number\ of\ transport\ blocks\ per\ T\ TI}{T\ TI} \times transport\ block\ payload\ size \qquad (2.5)$$

All the parameters used in this formula can be derived from the transport format definition used in NBAP signalling messages. To understand this it is necessary to look at some selected parameters in message example 2.12. Although this is just an excerpt of the full NBAP message and most sequences have been hidden it is still a fairly huge signalling portion, but is essential to understand how throughput is computed.

Message example 2.12 Transport format sets in NBAP synchronised radio link reconfiguration request message

ID Name	Comment or Value	
TS 25.433 V3.7.0 (2001-09) (NBAP) initiatingMessage (= initiatingMessage)		
nbapPDU		
1 initiatingMessage		
1.1 procedureID		
1.1.1 **procedureCode**	id-synchronisedRadioLink-ReconfigurationPreparation	
1.5.1.4.1 id	id-FDD-DCHs-to-Modify	
1.5.1.4.3.1.4 dCH-SpecificInformationList		
1.5.1.4.3.1.4.1 dCH-ModifySpecificItem-FDD		
1.5.1.4.3.1.4.1.1 **dCH-ID**	31	
1.5.1.4.3.1.4.1.2 **ul-TransportFormatSet**		
1.5.1.4.3.1.4.1.2.1 dynamicParts		
1.5.1.4.3.1.4.1.2.1.1 sequence		
1.5.1.4.3.1.4.1.2.1.1.1 **nrOfTransportBlocks**	0	
1.5.1.4.3.1.4.1.2.1.1.2 mode		
1.5.1.4.3.1.4.1.2.1.1.2.1 notApplicable	0	
1.5.1.4.3.1.4.1.2.1.2 sequence		
1.5.1.4.3.1.4.1.2.1.2.1 **nrOfTransportBlocks**	1	
1.5.1.4.3.1.4.1.2.1.2.2 **transportBlockSize**	148	
1.5.1.4.3.1.4.1.2.1.2.3 mode		
1.5.1.4.3.1.4.1.2.1.2.3.1 notApplicable	0	
1.5.1.4.3.1.4.1.2.2 semi-staticPart		
1.5.1.4.3.1.4.1.2.2.1 **transmissionTimeInterval**	msec-40	
1.5.1.4.3.1.4.1.3 **dl-TransportFormatSet**		
1.5.1.4.3.1.4.1.3.1 dynamicParts		
1.5.1.4.3.1.4.1.3.1.1 sequence		
1.5.1.4.3.1.4.1.3.1.1.1 **nrOfTransportBlocks**	0	
1.5.1.4.3.1.4.1.3.1.1.2 mode		
1.5.1.4.3.1.4.1.3.1.1.2.1 notApplicable	0	
1.5.1.4.3.1.4.1.3.1.2 sequence		
1.5.1.4.3.1.4.1.3.1.2.1 **nrOfTransportBlocks**	1	
1.5.1.4.3.1.4.1.3.1.2.2 **transportBlockSize**	148	
1.5.1.4.3.1.4.1.3.1.2.3 mode		
1.5.1.4.3.1.4.1.3.1.2.3.1 notApplicable	0	
1.5.1.4.3.1.4.1.3.2 semi-staticPart		
1.5.1.4.3.1.4.1.3.2.**1 transmissionTimeInterval**	msec-40	
1.5.1.5.1 id	id-DCHs-to-Add-FDD	
1.5.1.5.3.1.5 dCH-SpecificInformationList		

Message example 2.12 (*Continued*)

ID Name	Comment or Value	
1.5.1.5.3.1.5.1 dCH-Specific-FDD-Item		
1.5.1.5.3.1.5.1.1 **dCH-ID**	24	
1.5.1.5.3.1.5.1.2 **ul-TransportFormatSet**		
1.5.1.5.3.1.5.1.2.1 dynamicParts		
1.5.1.5.3.1.5.1.2.1.1 sequence		
1.5.1.5.3.1.5.1.2.1.1.1 **nrOfTransportBlocks**	0	
1.5.1.5.3.1.5.1.2.1.1.2 mode		
1.5.1.5.3.1.5.1.2.1.1.2.1 notApplicable	0	
1.5.1.5.3.1.5.1.2.1.2 sequence		
1.5.1.5.3.1.5.1.2.1.2.1 **nrOfTransportBlocks**	1	
1.5.1.5.3.1.5.1.2.1.2.2 **transportBlockSize**	336	
1.5.1.5.3.1.5.1.2.1.2.3 mode		
1.5.1.5.3.1.5.1.2.1.2.3.1 notApplicable	0	
1.5.1.5.3.1.5.1.2.1.3 sequence		
1.5.1.5.3.1.5.1.2.1.3.1 **nrOfTransportBlocks**	2	
1.5.1.5.3.1.5.1.2.1.3.2 **transportBlockSize**	336	
1.5.1.5.3.1.5.1.2.1.3.3 mode		
1.5.1.5.3.1.5.1.2.1.3.3.1 notApplicable	0	
1.5.1.5.3.1.5.1.2.1.4 sequence		
1.5.1.5.3.1.5.1.2.1.4.1 **nrOfTransportBlocks**	3	
1.5.1.5.3.1.5.1.2.1.4.2 **transportBlockSize**	336	
1.5.1.5.3.1.5.1.2.1.4.3 mode		
1.5.1.5.3.1.5.1.2.1.4.3.1 notApplicable	0	
1.5.1.5.3.1.5.1.2.1.5 sequence		
1.5.1.5.3.1.5.1.2.1.5.1 **nrOfTransportBlocks**	4	
1.5.1.5.3.1.5.1.2.1.5.2 **transportBlockSize**	336	
1.5.1.5.3.1.5.1.2.1.5.3 mode		
1.5.1.5.3.1.5.1.2.1.5.3.1 notApplicable	0	
1.5.1.5.3.1.5.1.2.2 semi-staticPart		
1.5.1.5.3.1.5.1.2.2.1 **transmissionTimeInterval**	msec-20	
1.5.1.5.3.1.5.1.3 **dl-TransportFormatSet**		
1.5.1.5.3.1.5.1.3.1 dynamicParts		
1.5.1.5.3.1.5.1.3.1.1 sequence		
1.5.1.5.3.1.5.1.3.1.1.1 **nrOfTransportBlocks**	0	
1.5.1.5.3.1.5.1.3.1.1.2 mode		
1.5.1.5.3.1.5.1.3.1.1.2.1 notApplicable	0	
1.5.1.5.3.1.5.1.3.1.2 sequence		
1.5.1.5.3.1.5.1.3.1.2.1 **nrOfTransportBlocks**	1	
1.5.1.5.3.1.5.1.3.1.2.2 **transportBlockSize**	336	
1.5.1.5.3.1.5.1.3.1.2.3 mode		
1.5.1.5.3.1.5.1.3.1.2.3.1 notApplicable	0	
1.5.1.5.3.1.5.1.3.1.3 sequence		
1.5.1.5.3.1.5.1.3.1.3.1 **nrOfTransportBlocks**	2	
1.5.1.5.3.1.5.1.3.1.3.2 **transportBlockSize**	336	
1.5.1.5.3.1.5.1.3.1.3.3 mode		
1.5.1.5.3.1.5.1.3.1.3.3.1 notApplicable	0	
1.5.1.5.3.1.5.1.3.1.4 sequence		

Message example 2.12 (*Continued*)

ID Name	Comment or Value	
1.5.1.5.3.1.5.1.3.1.4.1 **nrOfTransportBlocks**	3	
1.5.1.5.3.1.5.1.3.1.4.2 **transportBlockSize**	336	
1.5.1.5.3.1.5.1.3.1.4.3 mode		
1.5.1.5.3.1.5.1.3.1.4.3.1 notApplicable	0	
1.5.1.5.3.1.5.1.3.1.5 sequence		
1.5.1.5.3.1.5.1.3.1.5.1 **nrOfTransportBlocks**	4	
1.5.1.5.3.1.5.1.3.1.5.2 **transportBlockSize**	336	
1.5.1.5.3.1.5.1.3.1.5.3 mode		
1.5.1.5.3.1.5.1.3.1.5.3.1 notApplicable	0	
1.5.1.5.3.1.5.1.3.2 semi-staticPart		
1.5.1.5.3.1.5.1.3.2.**1 transmissionTimeInterval**	**msec-20**	

Definitions for two different transport channels are found in this message example. Since uplink and downlink transmission on the radio interface require separate channels there are four DCHs on the Uu interface using the same DCH-ID, but different uplink or downlink transport format sets. On the Iub interface it is also necessary to measure uplink and downlink throughput separately, but here uplink and downlink is indicated by the direction indicator information element in the Iub frame protocol (FP, 3GPP 25.427).

The DCH with DCH-ID = 31 obviously already existed before the NBAP synchronised radio link reconfiguration preparation procedure is requested, because it is found in the FDD-DCHs-to-Modify sequence while DCH with DCH-ID = 24 is added to the connection.

In DCH 31, in uplink as well as in downlink, either zero transport blocks or one transport block of 148 bits is transmitted within a 40-millisecond time transmission interval (TTI). The specification of zero transport blocks is necessary to ensure the synchronisation of UTRAN data transmission. In this so-called *silent mode* FP entities only count TTIs using internal timers to stay in synchronisation. If calculating the maximum theoretical throughput of this channel by using these values the following result occurs:

$$\frac{1 \ (block) \times 148 \ bit}{40 \ msec} = 3.7 \ kbps \qquad (2.6)$$

If we looked at the appropriate RRC signalling we would also be able to see that SRBs are mapped onto this transport channel. Now looking at 3GPP 34.108 *Common Test Environments for User Equipment (UE) Conformance Testing* (a specification that defines all transport formats to be supported by UE software) proves that no 3.7 kbps DCCHs for SRBs are defined in this document, only standalone SRBs with 3.4 kbps, which are also used as associated SRBs in combination with CS and PS RABs (see Table 2.17), are defined.

Table 2.17 shows the transport format parameters found for DCH 31 in NBAP signalling. Field sizes and maximum data rates of payload in bps can be found the in the RLC section of the table.

The situation for DCH-ID = 24 is similar. Zero, one, two, three or four transport blocks of 336 bits can be transmitted on this dedicated transport channel within a time transmission

Table 2.17 Transport channel parameters for 3.4 kbps SRBs for DCCH (from 3GPP 34.108)

Higher layer	RAB/signalling RB User of radio bearer	SRB#1 RRC	SRB#2 RRC	SRB#3 NAS_DT High priority	SRB#4 NAS_DT Low priority
RLC	Logical channel type	DCCH	DCCH	DCCH	DCCH
	RLC mode	UM	AM	AM	AM
	Payload sizes, bits	**136**	**128**	**128**	**128**
	Max data rate, bps	**3400**	**3200**	**3200**	**3200**
	AMD/UMD PDU header, bit	8	16	16	16
MAC	MAC header, bit	4	4	4	4
	MAC multiplexing	4 logical channel multiplexing			
Layer 1	TrCH type	DCH			
	TB sizes, bit	148 (alt. 0,148) (note)			
	TFS TF0, bits	**0 × 148 (alt. 1 × 0) (note)**			
	TF1, bits	**1 × 148**			
	TTI, ms	**40**			
	Coding type	CC 1/3			
	CRC, bits	16			
	Max. number of bits/TTI before rate matching	516			
	Max. number of bits/radio frame before rate matching	129			
	RM attribute	155 to 165			

interval of 20 milliseconds. Using these values the theoretically maximum data rate would be:

$$\frac{4\,(blocks) \times 336\,bit}{20\,msec} = 67.2\,kbps \qquad (2.7)$$

Due to the difference between the block size and the payload field size of the transport block we get a higher data rate than provided for user data. Indeed the RLC payload size of these transport blocks is 320 bits (the remaining 16 bits are used for the RLC header). Using this correction the maximum RLC data rate is as shown Table 2.18

$$\frac{4\,(blocks) \times 320\,bit}{20\,msec} = 64\,kbps \qquad (2.8)$$

Using these tables found in 3GPP 34.108 it seems to be relatively easy to implement transport channel throughput measurement in protocol-analysis based performance measurement software. It looks like the software just needs to detect the transport format parameters of the transport format set in NBAP signalling belonging to a single UE connection. Then an internal mapping table provides payload sizes and a filtering dimension can be constructed to distinguish between measurements on different radio bearer types (remember that radio

Table 2.18 Transport channel parameters for interactive or background 64 kbps
PS RAB (from 3GPP 34.108)

Higher layer	RAB/Signalling RB	RAB
RLC	Logical channel type	DTCH
	RLC mode	AM
	Payload sizes, bits	**320 (alt. 128)**
	Max data rate, bps	**64 000**
	AMD PDU header, bit	16
	MAC header, bits	0
MAC	MAC multiplexing	N/A
Layer 1	TrCH type	DCH
	TB sizes, bits	336 (alt. 144)
	T TF0, bits	**0×336 (alt. 0×144)**
	F TF1, bits	**1×336 (alt. 1×144)**
	S TF2, bits	**2×336 (alt. 3×144)**
	TF3, bits	**3×336 (alt. 7×144)**
	TF4, bits	**4×336 (alt. 10×144)**
	TTI, ms	**20**
	Coding type	TC
	CRC, bits	16
	Max. number of bits/TTI after channel coding	4 236 (alt. 4 812)
	Max. number of bits/radio frame before rate matching	2 118 (alt. 2 406)
	RM attribute	130 to 170

bearer data rates may change during an active connection due to changing radio transmission
conditions or traffic volume measurements while RAB for 64 kpbs remains the same).

But there are exceptions, of course. If segments belonging to different RLC service data
units (SDUs) are transmitted in the same RLC transport block the end of the first SDU and
the beginning of new SDU need to be marked using additional RLC signalling information,
which reduces the amount of payload bits in the transport block. The same happens if the
block only contains the last segment of a SDU plus padding bits that do not have any
information content and are used to fill up empty space in signalling messages and transport blocks. In contrast to RLC header portions padding bits need to be included in transport
channel throughput measurement. No matter if the RLC header is excluded or included the
real maximal throughput will never exactly meet the theoretical maximal throughput. It will
always be higher (if the RLC header included) or lower (if header excluded).

*Note: if a RLC transport block contains the last fragment of the first SDU and the first
fragment of the next SDU the payload amount is one full byte less. If a RLC block contains
the last fragment of an SDU plus padding bits the payload amount is two bytes less.*

In Figure 2.16 such a transport block containing the last data segment of a SDU plus
padding bits is shown. The bitmask column shows the length of single information elements
where 'b' stands for bit and 'B' for byte. As can be seen the last data segment of SDU is 9 bytes
long and 29 bytes of padding follow, which is in total a payload field size of $29 + 9 = 38$ bytes.

ID Name	Comment or Value	BITMASK
2.2 FP: Transport Block		
2.2.1 MAC: Target Channel Type	DTCH (Dedicated Traffic Channel)	
2.2.2 MAC: RLC Mode	Acknowledge Mode	
2.2.3 RLC: Data/Control	Acknowledged mode data PDU	1-------
2.2.4 RLC: Sequence Number	84	**b12***
2.2.5 RLC: Polling Bit	Status report not requested	-----0--
2.2.6 RLC: Header extension type	Octet contains LI and E bit	------01
2.2.7 RLC: Length Indicator	9	0001001-
2.2.8 RLC: Extension Bit	The next field is LI and E bit	-------1
2.2.9 RLC: Length Indicator	Rest is padding	1111111-
2.2.10 RLC: Extension Bit	The next field is data	-------0
2.2.11 RLC: Last Data Segment	03 63 6f 6d 00 00 01 00 01	***B9***
2.2.12 RLC: Padding	aa aa aa aa aa aa aa aa aa aa aa aa aa aa...	**B29***

Figure 2.16   RLC transport block with last data segment and padding

The missing two bytes ($= 6$ bits), compared to the payload size defined for this channel following Table 2.17, are used for two additional length indicators and extension bits.

Now with the necessary input to compute transport channel throughput all the performance measurement software needs to do is to count the number of transport blocks for a defined channel type in the uplink and downlink direction. Using a sampling period defines how often and when counters are reset and a computed measurement result is available. A perfect sampling period is one second, because it ensures optimum performance of throughput measurement application.

Counting of transport blocks must start with the first monitored FP DATA DCH frame on uplink or downlink. Figure 2.17 shows how throughput in the uplink direction is measured

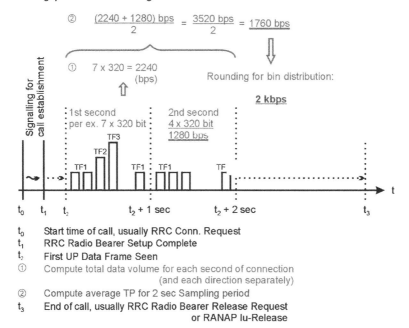

Figure 2.17   UL transport channel throughput measurement using 2-second sampling period and bin conversion

on a transport channel using a 2-second sampling period. Transport format numbers equal the number of transport blocks in FP frames: TF1 = 1 transport block in set, TF2 = 2 transport blocks in set and so on.

It is possible to define throughput bins to be able to compute a distribution function. Bin definition in such cases are user defined, e.g. bin 0 could represent 0 ... 0.999 kbps, bin 1 equals a range 1 ... 1.999 kbps and so on. Definition of bins leads to rounding throughput measurement results as also shown in Figure 2.17.

The transport channel throughput measurement is stopped when the RAB is deleted.

Using dynamically changing radio bearer types and UE RRC mobility tracking it is possible to correlate throughput measurement with (dynamically changing) radio bearer configurations and to compute throughput on cell level. Since transport channels are established between the SRNC and UE they are often transmitted using more than one radio link. Hence, macro-diversity filtering is necessary for uplink measurement and a similar filter method needs to be used for downlink traffic. However, there is no macro-diversity combining on downlink. Instead all RLC transport blocks are sent using multiple radio links on the radio interface, because the UE performs maximum ratio combining of all received radio signals. Therefore downlink data filtering on the Iub interface must be based on proprietary definitions of measurement equipment manufacturers. This downlink filtering mechanism could be called SSDT-likewise downlink filtering, because in the near future site selection diversity transmission (SSDT) will allow the transmission of downlink data using only the best cell of an active set. The best cell for downlink transmission is evaluated using feedback information (FBI) sent by the UE on the DPCCH in uplink. Once this feature is introduced it will result in call scenarios having several uplink data streams, but only one downlink data stream. The target of downlink data filtering is to ensure that each downlink transport block is counted only once although monitored on several Iub interfaces simultaneously, which is the same effect as in the SSDT scenario. Indeed, it makes sense to count downlink transport blocks on the link of the best cell that has been detected by performance measurement software evaluating radio quality indicators. Using these rules consequently for each uplink or downlink data frame the best cell used for transmission via Uu is known. Changes of the best cell are tracked by performance measurement software, but changes of the best cell bear the risk of throughput measurement failures if the best cell is changed in the middle of a sampling period (which is the rule) and throughput is aggregated on cell level. To illustrate this problem it is assumed that there is an ideal call with a constant 64 kbps throughput in all sampling periods as shown in Figure 2.18. The call is started in cell 1 and following a radio link addition cell 2 becomes the best cell. This change of best cell happens exactly in the middle of a sampling period of 2 seconds.

From the point of view of the UE the throughput is still at 64 kbps, but average values per cell 1 and per cell 2 for the sampling period with cell change will be lower. This occurs because the throughput is counted based on transport block payload size, the number of transport blocks and TTI, in given example 32 kbps for the last period of cell 1 and 32 kbps for first sampling period of cell 2. This is correct, because it shows real data transfer per time per cell in statistics. However, if the user does not know about the details of this measurement procedure a misinterpretation of measurement results is likely. Such a misinterpretation may lead to the wrong conclusion that not enough transport resources have been provided by the network.

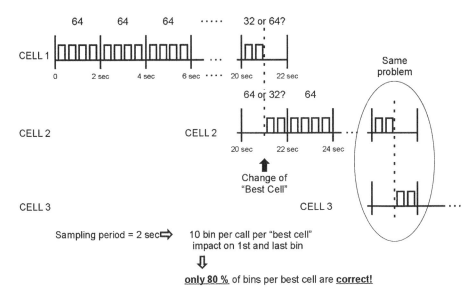

Figure 2.18   Cell throughput measurement failure due to cell change/handover

A possible solution to prevent this kind of problem is to summarise the measurement result, which means: if the best cell is changed after one second the measurement related to cell 1 will ignore the remaining second and provide the measurement result of 64 kbps for this whole sampling period. This rule must apply to both cells, cell 1 and cell 2, for the same sampling period. If measurement software summarises samples of single calls to calculate the throughput on different aggregation levels (e.g. per RNC/per RAB) the two summarised samples will be added and a total throughput of 128 kbps for this particular sampling period will be computed. The same thing happens again when cell change is performed from cell 2 into cell 3.

A third option is to summarise the sample for cell 1 only and completely ignore the frames sent via cell 2 for this sampling period. This will help to overcome problems of the first two cases, but requires more complicated measurement algorithms in software implementation. In addition, aggregation on the best cell level is no longer precise and if e.g. protocol events for soft handover are correlated to throughput measurement results this will show strange results in time evaluation analysis: although a cell becomes the best cell no data is transmitted on this new best cell during the first second after a successful best cell change.

These are the three possible options and whatever will be implemented in detail there will always be an advantage and a disadvantage combined with a measurement failure/tolerance. The longer the sampling period the higher the impact caused by measurement failures. In average the best cell is changed after 20 seconds due to soft or softer handover. Using a 2-second sampling period within 20 seconds, 10 throughput samples will be computed. Considering the first as well as the last sample to be erroneous it must be considered that only 80% of all samples show proper measurement results. Using a 1-second sampling period the total number of samples will be 20. Two of these samples are erroneous and hence, the absolute failure of measurement is decreased to 10%. Granularity limitation is the TTI value of transport channel.

It is also true that throughput measurement on an HS-DSCH is completely different from the procedure described in the previous paragraphs. HS-DSCHs work in a different way than DCHs. The main differences are:

- Several UEs/connections send their downlink data using the same HS-DSCH.
- Packet scheduling and multiplexing for HS-DSCH is done in Node B.
- Retransmissions on the radio interface are handled internally by the HARQ entities of UE and Node B and for this reason retransmitted RLC blocks on radio interface are not visible on Iub.
- The maximum downlink data rate on HS-DSCH depends not only on the capacity of the transport channel, but also on capabilities of UEs that are desribed using different HSDPA capability classes.

Note: there is no transport format defined for HS-DSCHs and hence it is not possible to compute the maximum downlink data rate on HS-DSCHs as is possible in the case of DCHs. Consequently, it is also not possible to define a radio bearer type using a specific downlink data rate.

What can be measured on Iub is the throughput perceived on different Iub physical transport bearers (AAL2 SVCs) carrying MAC-d flows of single UEs that are later multiplexed by Node B into the same HS-DSCH. However, due to the nature of radio interface transmission of HS-DSCH data as described in the previous paragraphs it is not possible to measure the transport channel throughput of HS-DSCHs precisely and the same statement is true if RLC throughput on HS-DSCHs needs to be measured.

Analysing how throughput samples measured on different calls/transport channels are summarised on cell aggregation level another problem occurs. This problem is easy to explain: if throughput is measured on cell level using e.g. a 2-second sampling period a call will never start exactly at the beginning of a sampling period. This leads to a displacement of measurement results, because usually measurement architecture first filters all frames on behalf of call trace application, then it computes throughput samples per call/channel and at the end it adds samples of all calls in one cell to compute throughput per cell. To compute throughput samples per call is executed with higher priority and therefore the first measurement sample of a single call is available if the first sampling period per call is finished. However, if this call-sampling period started in the middle of a cell-sampling period the cell-sampling period is already expired when the measurement result for call-sampling period becomes available. In other words, the real throughput in the cell during the cell-sampling period is higher than in the presented measurement result, because the measurement result of a specific call that has just started during this cell-sampling period is not available yet. Figure 2.19 illustrates this problem.

If it is assumed that each call has a constant transport format (the maximum data rate of a certain radio bearer type) and each defined transport format equals a certain RAB-Type (TF1 = RAB1, TF2 = RAB2 etc.), it is proven that the same displacement effect also interferes with measurement results of other dimensions, for instance RAB-Type and RB-Type.

The problem becomes clearly visible if we look at the first sampling period of the cell. TF1 represents 32 kbps, TF2 equals 64 kbps, TF3 = 128 kbps and TF4 = 384 kbps. Total throughput of the first sampling period in the cell is the total value from measurement results

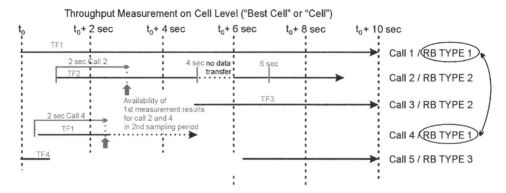

Figure 2.19 Cell throughput measurement failure due to call sample addition

for call $1 = 32$ kbps plus call $5 = 384$ kbps ($384 + 32 = 416$ kbps). But in fact there has also been 50% of data of a TF2 $= 64$ kbps sample and 75% of another TF1 (32 kbps) in the same sampling period. This means that the real cell throughput for this sampling period was actually 0.5×64 kbps $+ 0.75 \times 32$ kbps 416 kbps $= 472$ kbps. These are 56 kbps more than shown in measurement results. In turn the missed 56 kbps of the first sample is shown as additional load in the second sampling period. By using a shorter sampling period this measurement failure will become less, but it will never completely disappear.

Once again this problem could be solved by the implementation of a cell trace application, but this cell trace should work after call trace, because call trace is a mandatory prerequisite for macro-diversity uplink filtering and SSDT-likewise downlink filtering of data streams. This is certainly not an easy design task and again it should be considered carefully if an increased precision of measurement results is in proportion to the necessary development effort.

2.3.3 PACKET SWITCHED USER PERCEIVED THROUGHPUT

User perceived throughput means to measure throughput on the RLC payload level excluding padding bits and retransmitted transport blocks. The easiest way to filter out padding and retransmitted RLC blocks is to look at the layer on top of the RLC, which is the internet protocol (IP) level of the UTRAN user plane. On this IP layer there are not only IP frames, but also a suite of the following IP versions:

- IPv4 – Internet Protocol version 4 (32-bit addresses)
- IPv6 – Internet Protocol version 6 (128-bit addresses)
- PIP – The 'P'-Internet Protocol
- IP/ST – IP in ST Datagram Mode
- TP/IX – The 'Next' Internet Protocol
- TUBA – TCP/UDP over Connectionless-mode Network Layer Protocol (CLNP)

To discuss the differences between these IP versions is beyond the scope of this book, however, they are discussed in one of the publications about IP services in 3G mobile networks.

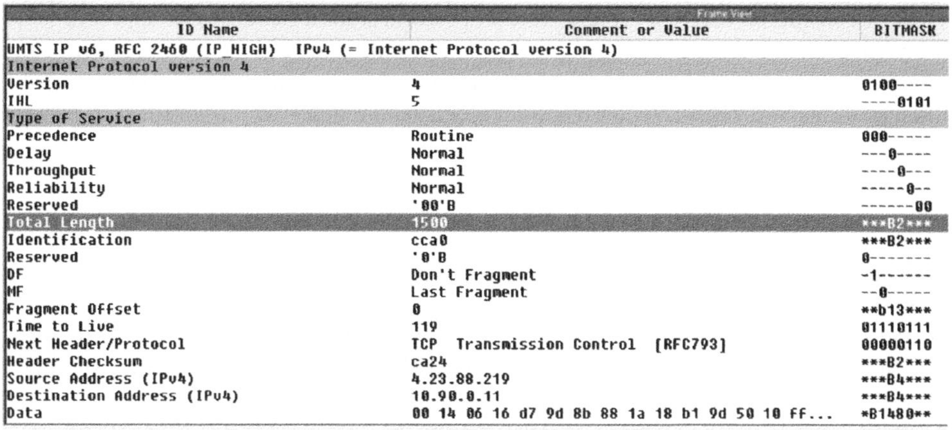

```
                                                                    Frame View
              ID Name                                    Comment or Value                      BITMASK
UMTS IP v6, RFC 2460 (IP_HIGH)   IPv4 (= Internet Protocol version 4)
Internet Protocol version 4
Version                          4                                                             0100----
IHL                              5                                                             ----0101
Type of Service
Precedence                       Routine                                                       000-----
Delay                            Normal                                                        ---0----
Throughput                       Normal                                                        ----0---
Reliability                      Normal                                                        -----0--
Reserved                         '00'B                                                         ------00
Total Length                     1500                                                          ***B2***
Identification                   cca0                                                          ***B2***
Reserved                         '0'B                                                          0-------
DF                               Don't Fragment                                                -1------
MF                               Last Fragment                                                 --0-----
Fragment Offset                  0                                                             **b13***
Time to Live                     119                                                           01110111
Next Header/Protocol             TCP   Transmission Control   [RFC793]                         00000110
Header Checksum                  ca24                                                          ***B2***
Source Address (IPv4)            4.23.88.219                                                   ***B4***
Destination Address (IPv4)       10.90.0.11                                                    ***B4***
Data                             00 14 06 16 d7 9d 8b 88 1a 18 b1 9d 50 10 FF...               *B1480**
```

Figure 2.20 IP frame header

To ensure a correct measurement result it is only important that protocol frames on this layer are correctly decoded to be usable for user-perceived throughput measurement purposes.

There are two different points at which the throughput of the user plane IP can be measured: the Iub/Iur and the IuPS interface. On IuPS IP frames are transported unciphered and embedded in the GPRS tunnelling protocol (GTP) T-PDUs (T stands for tunnel). The payload field of a single T-PDU equals the content of a whole RLC SDU. Hence, it is quite easy to measure user-perceived throughput on IuPS. The only difficulty for measurement is distinguishing between uplink and downlink IP frames, but this can be achieved if the appropriate GTP tunnel endpoint identifiers (TEIDs) for the SRNC and SGSN are stored as a context in the software as long as the tunnel is active. Those TEIDs are negotiated in the RANAP RAB assignment procedure in IuPS.

The volume of IP data (RLC SDUs) can easily be measured by using the length field of the IP header, which indicates the size of the total IP frame: header plus data. An example is given in Figure 2.20. Here the frame has a total length of 1480 byte data + 20 byte header = 1500 byte.

The only problem when measuring user-perceived payload on IuPS is the correlation with signalling on Iub and Uu. Take a look at the delay between the arrival of a downlink IP frame on the SRNC and the same frame reassembled on Iub after the last RLC transport block has been sent (time stamp of reassembled frame uses time stamp of last transport block) as shown in Figure 2.21. It becomes apparent that the delay between these two interfaces (anchor symbol: IuPS, arrow symbol: Iub) is more than 844 ms, which is quite close to one second.

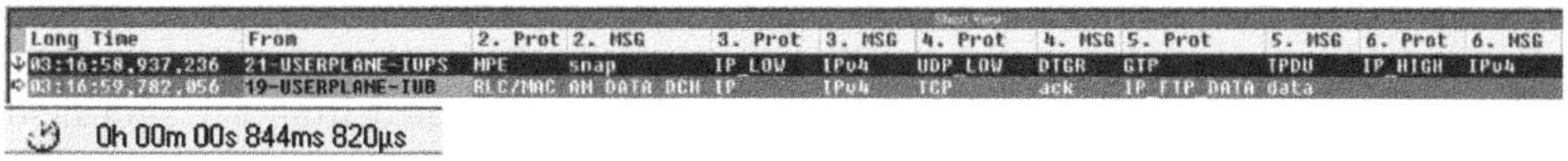

```
                                                              Short View
 Long Time              From          2. Prot 2. MSG   3. Prot  3. MSG  4. Prot  4. MSG 5. Prot   5. MSG  6. Prot  6. MSG
 03:16:58,937,236   21-USERPLANE-IUPS HPE     snap     IP_LOW  IPv4     UDP_LOW  DTGR   GTP        TPDU    IP_HIGH  IPv4
 03:16:59,782,056   19-USERPLANE-IUB  RLC/MAC AM DATA DCH IP   IPv4     TCP      ack    IP_FIP DATA data

   Oh 00m 00s 844ms 820µs
```

Figure 2.21   Transmission delay of IP frame between IuPS and Iub

If it is considered that the typical time to perform a hard handover is approximately 200 milliseconds it means that – depending on uplink or downlink measurement – the software would count the throughput on the target cell of handover already, although it has still been sent on the source cell or the other way round. Since it must be kept in mind that a single RLC SDU consists of up to 12 000 bits is makes a difference on which cell or in which sampling period data volume of this SDU have been counted.

In turn it is easy to correlate RLC SDUs reassembled on Iub to single transport blocks that have been used to transmit them and it is also easy to identify with high precision on which cell or active set these transport blocks have been sent/received at a certain time.

Conclusion: as long as the user plane throughput of a single call only is to be measured it makes no difference on which interface (Iub, IuPS, Gn, Gi) this measurement is done. However, if the measurement result is to be correlated to the location of the UE at a certain time or to radio-specific quality parameters like BER or BLER it is necessary to measure user-perceived throughput on Iub to be as close to the radio interface as possible.

2.3.4 APPLICATION THROUGHPUT

Application throughput is the data volume measured on the layer 7 protocol of the TCP/IP user plane protocol stacks divided by the time taken to transit this amount of data. For this type of throughput measurement the same requirements are valid as mentioned in the case of user-perceived throughput: any correlation with the radio interface topology and signalling requires measurement on Iub.

For application throughput measurement the challenge is not decoding, reassembly and filtering of data, it is finding out about the application data volume itself. Look for example at an FTP frame as presented in Figure 2.22.

This FTP frame contains pure data and there is no length field in TCP but a data offset information field. This data offset is related to the first bit of the TCP frame and indicates where the payload field of the TCP frame begins. Since the whole TCP frame is organised in multiples of 32 bits (4 bytes) it is clear that 5×4 bytes $= 20$ bytes need to be subtracted from the total TCP frame length to get the payload field size. Now the problem is that the previous IP frame that carries TCP does not contain any information about its payload size, only the total length of the IP frame including header. Often the IP header is 20 bytes long, but we cannot count on this fixed value, because due to a number of options IP headers may become extremely long – so long that special algorithms for IP header compression have been defined.

To calculate FTP data volume using information in the bitmask column would be easy, but bitmask information is delivered by the decoder of the measurement unit, not by the protocol itself. To refer to this decoder-specific information it needs a more sophisticated measurement application than one that just adds up values of length fields.

For the user datagram protocol (UDP) it might be a little bit easier, because the UDP header has a fixed size of 8 bytes, which can be subtracted from a length field value that indicates the size of the whole UDP frame: header plus UDP contents (see Figure 2.23).

ID Name	Comment or Value	BITMASK
Delay	Normal	---0----
Throughput	Normal	----0---
Reliability	Normal	-----0--
Reserved	'00'B	------00
Total Length	**1500**	***B2***
Identification	cca0	***B2***
Reserved	'0'B	0-------
DF	Don't Fragment	-1------
MF	Last Fragment	--0-----
Fragment Offset	0	**b13***
Time to Live	119	01110111
Next Header/Protocol	TCP Transmission Control [RFC793]	00000110
Header Checksum	ca24	***B2***
Source Address (IPv4)	4.23.88.219	***B4***
Destination Address (IPv4)	10.90.0.11	***B4***
Data	00 14 06 16 d7 9d 8b 88 1a 18 b1 9d 50 10 ff...	*B140**
TCP – Transmission Control Protocol, RFC793 (+RFC1072/1323/2... (TCP) ack (= acknowledge)		
acknowledge		
Source Port	ftp-data – File Transfer [Default Data]	***B2***
Destination Port	xingmpeg	***B2***
Sequence Number	3617426312	***B4***
Acknowledgment Number	437825949	***B4***
Data Offset	5	0101----
Reserved	'000000'B	***b6***
URG	Urgent Pointer field not significant	--0-----
ACK	Acknowledgement field significant	---1----
PSH	No Push Function	----0---
RST	Don't reset the connection	-----0--
SYN	Don't synchronize sequence numbers	------0-
FIN	More data from sender	-------0
Window	65535	***B2***
Checksum	e6b5	***B2***
Urgent Pointer	0	***B2***
data	91 e8 3f ff 9d e3 bf a0 d0 06 60 00 b6 10 00...	*B140**
FTP-DATA – File Transfer Protocol (Data), RFC959 (IP_FTP_DATA) data (= Data)		
Data		
Data	91 e8 3f ff 9d e3 bf a0 d0 06 60 00 b6 10 00...	*B140**

Figure 2.22   TCP and FTP data frame

ID Name	Comment or Value	BITMASK
UDP, RFC 768 08.80 (UDP_HIGH) DTGR (= Datagram)		
Datagram		
Source Port	– unknown / undefined –	***B2***
Destination Port	Domain Name Server (DNS)	***B2***
Length	**69**	***B2***
Checksum	32203	***B2***
UDP contents	1a 10 01 00 00 01 00 00 00 00 00 00 09 75 73...	**B61***
DNS – Domain Name Service (RFC1035) (DNS) qry (= query)		
query		
header		
id	1a10	***B2***
qr	query	0-------
opcode	standard query	-0000---

Figure 2.23   UDP datagram with length indicator

2.4 TRANSPORT CHANNEL USAGE RATIO

The transport channel usage ratio is a simple formula that describes the relation between maximum possible data rates provided by transport channels and the average throughput measured when a channel has been active. This time is called the observation period in the transport channel usage ratio formula shown below:

$$\frac{DataVolume\ of\ RLC\ TransportBlocks\ (kbit)}{max\ Theoretical\ transport\ channel\ throughput\ (kbit/s) \times observation\ period(s)} \times 100\%$$

(2.9)

Table 2.19   Example of transport channel usage ratio analysis in a certain SRNC

	UL transport channel usage ratio	DL transport channel usage ratio
Cell 1		
RB-Type UL/DL 64/64 kbps	60%	85%
RB-Type UL/DL 64/128 kbps	25%	80%
RB-Type UL/DL 64/384 kbps	25%	90%

Using this formula a good ratio can be computed that describes how often (in %) it has been necessary to use the maximum bandwidth of a single transport channel. Note that this analysis must be done separately for uplink and downlink traffic.

For a single call analysis it might be enough to know the transport channel ID. Usually the same transport channel ID is used to identify a DCH even if the call was in CELL_FACH state for a period of time and all dedicated transport channels have been deleted. However, if the analysis of multiple connections need to be performed on a certain aggregation level such as cell or SRNC it is necessary to know the radio bearer type of the DCHs (uplink/downlink maximum bit rate on the radio interface following definitions of 3GPP 34.108). This is because the same channels having the same DCH-ID may have different transport format settings and only measurement results from channels of the same type/transport format can be used to calculate an average of multiple measurements. Here it must also be kept in mind that this radio bearer type assigned to a certain DCH by the call admission control function of SRNC may change during the call due to dynamical reconfiguration as described in section 1.1.6.

Table 2.19 shows an example of such an analysis for a single cell. It makes sense to aggregate measurements of multiple calls using UL/DL combinations of maximum bit rates to gain a better overview of how single connections of the same type behave. A low ratio value for a certain maximum data rate does not necessarily indicate a problem, it just means that not much data has been transmitted on these UL DCHs, for example because the used application service is a downlink file transfer (in this case only TCP Acknowledgements and RLC AM Status PDUs over a long time on UL are recognised). Another aspect is that on uplink a low transport channel usage ratio must not be seen as critical, because the number of uplink channelisation codes is not a critical factor, as each uplink dedicated physical channel is uniquely identified by an uplink scrambling code. Channelisation codes on uplink are only necessary to distinguish between DPCCH and DPDCH of the same connection. The situation on downlink is different. Here a single cell uses the same downlink scrambling code for all connections, hence the channelisation code is a unique identifier of a single connection. But the number of available channelisation codes per cell is limited due to the structure of the channelisation code tree. Also the number of available codes determines the capacity limit of the cell. To give an example, there are only four channelisation codes with spreading factor 4, which represents the highest downlink data rate possible for a single connection on the radio interface in non-HSDPA capable cells. If these four available spreading codes are used simultaneously no other connections are possible using the same cell. The spreading factor is indirectly (using the coding rate of the convolutional coder) related to the downlink maximum bit rate of the radio bearer, which represents the maximum theoretical transport channel throughput used to compute the transport channel usage ratio. If the transport channel usage ratio of channels with the highest available downlink bit rate is low, this means that these transport resources are not efficiently administrated by the SRNC and high-speed transport resources are blocked by users that need much lower bit rates, but for lower bit rates higher spreading factors are used. And the higher the spreading factor the more connections can be served in a cell at the same downlink bit rate.

If a cell is HSDPA capable the transport channel usage ratio of HS-DSCH cannot be computed, because it is very difficult to estimate the theoretical maximum throughput of an HS-DSCH. Large parts of data volume transported on this channel are hidden when monitoring the Iub interface. This is because retransmissions on the radio interface due to hybrid ARQ (HARQ) error detection/correction cannot be monitored on Iub, but will most likely require a huge portion of HS-DSCH transport capabilities.

If the data volume of RLC transport blocks is counted, analysis algorithms run again into a problem discussed in the section about transport channel throughput. As it was pointed out in this section an RLC transport block contains not only the payload to be transported, but also RLC/MAC header bits. If the transport channel throughput is measured related to a 64 kbps radio bearer using full transport block size of 336 bits, it must also calculate the maximum theoretical transport channel throughput using 336 bits. To base the measurement on a transport block size of 336 bits leads to a maximum value of 67.2 kbps for a 64 kbsp radio bearer. The advantage of this variant of measurement is that it reflects the real traffic situation on the DCH, which is also used to transport the RLC header and padding bits as well as retransmitted RLC frames and RLC signalling messages (status PDUs). The resulting volume of all these different kinds of data finally also determines which spreading factor needs to be chosen for radio interface transmission.

Some radio network planners have a different point of view of this KPI and say it must be based on the data volume of RLC SDUs instead of the data volume of RLC transport blocks as shown in Equation (2.10):

$$\frac{DataVolume \ of \ RLC \ SDUs \ (kbit)}{max \ Theoretical \ transport \ channel \ throughput \ (kbit/s) \times observation \ period(s)} \times 100\%$$

$$(2.10)$$

An explanation of this requirement is that the target of transport channel usage ratio analysis is to find out how well the RNC admission control function is able to assign the necessary transport resources for a particular payload bit stream. If the data volume of RLC SDUs is in the numerator the maximum theoretical transport channel throughput must be computed in a different way. Instead of the size of transport blocks it must be based on the available payload size of transport blocks. For transport channel throughput measurements it has already been demonstrated that the payload size of transport blocks is not a fixed value as assumed in 3GPP 34.108, but for a 336-bit transport block the payload size varies for single transport blocks between 38 bytes (304 bits) and 40 bytes (320 bits). If the maximum theoretical transport channel throughput is calculated using 320 bits, a maximum is defined that can never be reached. Hence, the maximum ratio calculated using this approach is not at 100%, but instead at 98 or 99%, which needs to be taken into account as a natural measurement tolerance. If the admission control function in SRNC software really uses user-perceived throughput (because this is what the data volume of SDUs stands for) to assign necessary radio resources to a connection is a question that can only be answered by network equipment manufacturers (NEMs) and this answer – if given at all – will always be a proprietary one.

Finally there is another aspect that may cause tolerance in measurement results in the range of up to approximately 10%: it is difficult to agree on a common definition when a transport channel is really available to transport data. Some measurement manufacturers use the ALCAP Establish Confirm message, which acknowledges the set up of an appropriate Iub/Iur physical transport bearer (AAL2 SVC); others wait until a number of FP synchronisation frames have been sent on the uplink/downlink of this AAL2 connection. However, there is no signalling message that indicates that the synchronisation process is finished. And in addition FP synchronisation frames are also sent as a check during long periods of DCH activity. A third option is to define the first user plane frame seen on AAL2 SVC as the start of availability, but in this case it is possible that in two directions (uplink or downlink) the first DCH frame will be seen quite late if there is no initial data to be transmitted in this direction. The time difference between ALCAP Establish Confirm and the first DCH frame is usually not more than a few hundred milliseconds. However, for PS connections with multiple channel type switching procedures these tolerance time frames also need to be multiplied, which leads to the fairly high measurement tolerance as mentioned at the beginning of this paragraph.

2.5 PRIMARY AND SECONDARY TRAFFIC

In Chapter 1 the benefits and disadvantages of soft handover in UMTS FDD mode were discussed. Now two KPIs are defined that measure the percentage of disadvantages: primary and secondary RLC traffic. As will be demonstrated these KPIs can only be calculated at cell level and a mandatory prerequisite is that during an active connection the performance measurement software always knows which cells belong to the active set of this connection.

Primary traffic is the traffic that is necessary to transmit data between the UE and the network and vice versa. If the UE is in a softer or soft handover situation identical RLC transport blocks are sent/received via each radio link belonging to the active set of the UE. One radio link is always related to a single cell. If there are e.g. two radio links related to two cells in an active set, double the amount of uplink and downlink RLC transport blocks is sent on the radio interface to benefit from the advantages of the soft handover scenario. One-half of this doubled traffic volume is called primary traffic, the other half is called the secondary traffic. It is important to understand that calculation of primary and secondary traffic follows straightforward statistical rules. There is no preference of a single radio link or a single cell in this formula.

Note: a logical failure that is often heard when discussing primary and secondary traffic is the assumption that all primary traffic is sent via the best cell of connection while all other links in the active set are only used to transmit secondary traffic.

The truth is that errors in radio transmission can occur on each link belonging to an active set. Hence, for the transmission of a certain RLC transport block it is not known which cell is the best cell to transmit this block error-free. If a cell offered such a high quality that transmission errors could be excluded the ongoing connection would not be in a soft handover situation. Following this statistically, primary and secondary traffic are seen as distributed to all radio links of the active set. The basic rule that is illustrated in Figure 2.24 says:

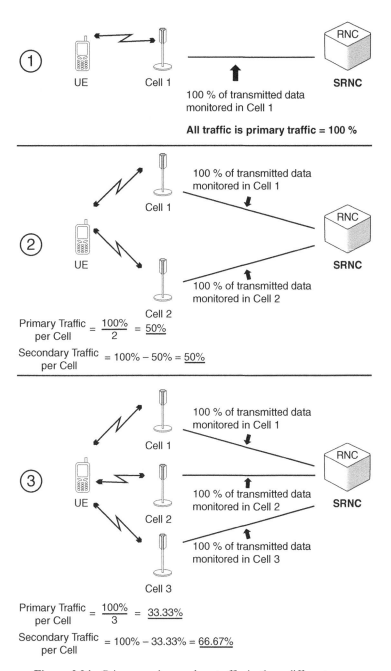

Figure 2.24 Primary and secondary traffic in three different cases

the more radio links a UE has in its active set the more secondary traffic and the less primary traffic is measured for this single UE connection.

To calculate primary and secondary traffic it is not necessary to measure the data volume or throughput of a connection. Looking at the different situations shown in Figure 2.24 it becomes clear that primary and secondary traffic as a percentage value can be calculated based on the average active set size of a single connection as shown in Equations (2.11) and (2.12). To aggregate primary and secondary traffic at a different aggregation level than the call (e.g. per cell or per SRNC) requires the average active set size of all connections that have been active using these network resources during a defined time period.

$$\text{Primary Traffic} = \frac{1}{\textit{Average Active Set Size of connection}} \times 100\% \qquad (2.11)$$

$$\text{Secondary Traffic} = 100\% - \text{Primary Traffic } (\%) \qquad (2.12)$$

It can be assumed that network operators would like to express how much radio transmission capacity is provided to transmit primary and secondary traffic, because minimising the secondary traffic component indicates the increase of the total capacity of the network. For this reason percentage values are combined with uplink and downlink RLC throughput measured on Iub/Iur physical transport bearers (AAL2 SVC) used to transport data streams of dedicated traffic channels (other channels are not involved when the UE is in soft handover). Although this calculation does not take into account the throughput caused by the FP header and trailer information bits it allows a good estimation of how much bandwidth is necessary to provide transport resources for secondary traffic in soft handover situations. For softer handover scenarios (cells involved in soft handover are located in the same Node B) this calculation does not apply, because usually Node B performs the combining of radio links and hence only a single AAL2 SVC is used on the Iub interface, but data is transmitted on two radio links in parallel.

Now the target is to find out how many additional resources are necessary to transmit secondary traffic. On wired interfaces, Iub and Iur, this is the bandwidth of electrical or optical cables, on the radio interface, Uu, bandwidth is correlated with the channelisation codes used in downlink. The number of available downlink channelisation codes limits the number of possible connections and services per cell.

Code utilisation per cell should rather be analysed by using a separate KPI, but transport overhead due to secondary traffic on Iub and Iur can be computed if the following prepositions are fulfilled:

- Performance measurement software must be able to distinguish between softer handover when diversity splitting/combining is done in Node B and soft handover situations when diversity splitting/combining is done in RNC.
- The overall measured transport channel throughput based on the size of transport blocks must be associated to each physical transport bearer of a single connection.
- Instead of the average active set size the average size of the used transport bearer set of connection must be known, which requires a special application that works in a similar way as the active set size tracking, but instead of the set up and deletion of

radio links per connection this other application must track the set up and deletion of VPI/VCI/CID per connection.

Secondary traffic on Iub/Iur physical transport resources could then be computed as shown in Equation (2.13):

$$\left(1 - \frac{1}{\textit{Average Size of TransportBearerSet}}\right) \times \text{Transport Channel Throughput [kbps]}$$

$$(2.13)$$

Whether the separation of uplink and downlink measurements is necessary may be decided for each particular case. In an ideal case the throughput on physical transport bearers can be measured at frame protocol level (including FP header and trailer information bits) and a topology detection tool provides additional information about which VPI/VCI/CID belong to which physical wire or fibre between network elements. Measurement results can be correlated with fixed costs of transport network resources, which especially in the case of leased lines can be expressed in costs per kbps bandwidth. The optimisation of secondary traffic can help to minimise these fixed costs. However, it must be kept in mind that the minimisation of secondary traffic is only possible as far as overall radio transmission is guaranteed and smooth soft handover procedures are guaranteed. Thus the optimisation of primary/secondary traffic is one of the most complex and most difficult procedures.

Another idea that is sometimes seen in requirement documents is the correlation of primary/secondary traffic percentage with transmitted data volume at the RLC SDU level, which is the level of user-perceived throughput. The objective behind such formulas is to estimate for how much traffic transmitted on UTRAN interfaces the user will be charged and how much 'payload' needs to be transmitted free of charge due to soft handover. Depending on how the fixed costs of network operation are distributed (e.g. per cell or per wire/fibre) the data volume of SDUs transmitted on uplink and downlink need to be measured and measurement values need to be distributed in the same way as cost (also per cell or per wire/fibre). If measurement results are computed at cell level, active set size tracking is necessary; at transport bearer level the same prerequisites apply as described for transport channel throughput.

The data volume of charged payload at cell level can be computed as shown in Equation (2.14):

$$\frac{\textit{DataVolume of UL} + \textit{DL RLC SDUs per cell}}{\textit{Average Active Set Size per cell}} \qquad (2.14)$$

The 'per cell' aggregation level in this formula is realised by calculating an average (active set size) or total value (data volume) of all connections that use dedicated radio links provided by this cell. There must be a high measurement tolerance taken into account due to fact that large SDUs are segmented into a huge number of transport blocks and the cell used for the transmission of these transport blocks can be changed with every transport block set that is sent or received.

2.6 ACTIVE SET SIZE DISTRIBUTION

The active set size distribution can be computed and displayed for each cell of the network. The purpose of this measurement is to find out how many radio links UEs have in their active set during the observation period of a full measurement session. Aggregation of measurement results for an unlimited number of individual calls shall be done at cell level. It can be computed and displayed how often UEs using a radio link of cell X have one, two, three or more radio links simultaneously in their active set.

The maximum active set size for UEs defined by 3GPP is six radio links. Due to capacity reasons most RNCs in today's networks are configured to allow a maximum of three radio links in the active set. In addition, 3GPP 25.942 *RF System Scenarios* refers to a number of laboratory simulations and tests and summarises test results as follows: '...assuming ideal handover measurements by UE and delay free handover procedure the gain of having more than 3 best cells in the active set is minimal. Thus, including extreme cases it can be concluded that UE does not have to support more than 4–6 as the maximum size of the active set.' Following this for the active set size distribution per cell it can also be assumed that the maximum number of radio links in the active set is six. The measurement can be displayed as a percentage value in a table in which each column represents a different active set size. An applicable graphic presentation format of this measurement is the histogram. Two case studies will help to explain how active set size distribution is calculated.

Case study 1: it is assumed that during the whole observation period of a monitoring session two UEs are served by a network cell. One of these UEs is in soft handover with a neighbour cell (active set size = 2) while the other UE is not in soft handover and has a single dedicated radio link in use provided by the cell for which the active set size distribution shall be computed. In this case the sample table would look as shown in Table 2.20.

However, at any time new radio links can be added to or existing radio links can be deleted from active sets of UEs. Knowing that this occurs Table 2.20 only depicts a momentary situation in a cell. To analyse active set sizes used by all UEs in a cell at any time it is necessary to take samples, but in contrast to BLER samples (discussed in section 2.1) there is no sampling period defined for this measurement and hence there is no reset of counter values if the sampling period timer expires. For the active set size distribution the first sample is taken at the beginning and the last sample is taken at the end of a monitoring session. The total number of samples depends on session duration.

Now another problem arises: which protocol events will be counted? The first idea is to look for NBAP radio link setup and radio link addition procedures, especially since Request messages of these procedures contain a radio link ID information element. The first radio link setup within an active set is indicated by Radio link ID = 0, the second one by Radio link ID = 1 and so on... until the first radio link is deleted from the active set while the link identified by Radio link ID = 1 remains in service. The next radio link to be added to the active set will be identified by Radio link ID = 0. And following this it becomes evident that the NBAP radio link ID cannot be used to distinguish how many radio links belong to the

Table 2.20   Active set size distribution example 1

Active set size	1	2	3	4	5	6
Average percentage	50%	50%	—	—	—	—

current active set of a single UE. Hence, a so-called invisible background application must run in performance measurement software that stores and tracks the changing active set size of each UE monitored in network. From this invisible background application the active set size of each UE can be polled periodically. In addition to the size of the active set the application must also know the identity of cells involved in the active set at any time.

The more samples polled during the observation period the higher the precision of measurement values. Thinking about the fact that the average usage time of a radio link in an active set in an optimised network is between 15 and 20 seconds it can be assumed that a polling period of one or two seconds provides a good minimum granularity for an active set size distribution measurement.

Case study 2: in the following example the session duration is 10 minutes and the sampling period is 2 seconds. As a result, it is expected to get a total number of 300 samples within the observation period (duration of monitoring session). If there is no UE active in the monitored cell the '0' counter values will be ignored.

Note: UEs that are in CELL_FACH state using common transport channels RACH and FACH of the monitored cell do not have a dedicated radio link in the active set. Hence, the active set size of UEs in CELL_FACH is zero.

It is assumed in this case study that within the observation period (10 minutes) three different UEs (represented by different line styles in Figure 2.25) have active dedicated radio links in this cell. UEs represented by dotted and dashed lines set up an initial dedicated radio link in the monitored cell, later go into soft handover with neighbour cells before the last radio link of the active set is deleted in the same cell in which the dedicated radio link has initially been established. The UE represented by a solid line enters and leaves the monitored cell following soft handover radio link addition and subsequent radio link deletion.

The active set size distribution percentage for each possible number of radio links in an active set is computed as the ratio of total counts of each possible active set size divided by the sum of all total counts as polled from the invisible background application multiplied by 100. If only the six samples shown as arrows in Figure 2.25 are taken into account, this will deliver the results shown in Table 2.21.

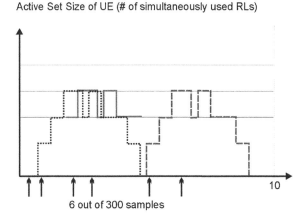

Active Set Size of UE (# of simultaneously used RLs)

6 out of 300 samples

Figure 2.25 Three UEs with changing active set size are monitored in a single UTRA cell

Table 2.21   Active set size distribution example 2

Active set size	1	2	3	4	5	6
Sample 1	Ignored, because zero UEs in cell					
Sample 2	1	0	0	0	0	0
Sample 3	0	1	1	0	0	0
Sample 4	0	0	2	0	0	0
Sample 5	1	0	0	0	0	0
Sample 6	0	0	1	0	0	0
Total counts	**2**	**1**	**4**	**0**	**0**	**0**
Average percentage	**28.6%**	**14.3%**	**57.1%**	—	—	—

The sum of total counts in Table 2.21 is seven. The sum of total counts in Table 2.21 is seven and calculation results are as follows:

$$\text{Average percentage active set size } 1 = (2/7) \times 100\% = \mathbf{28.6\%}$$
$$\text{Average percentage active set size } 2 = (1/7) \times 100\% = \mathbf{14.3\%}$$
$$\text{Average percentage active set size } 3 = (4/7) \times 100\% = \mathbf{57.1\%}$$

The histogram for this case study is shown in Figure 2.26.

In a real network environment the number of polls and the number of total counts is of course much higher due to longer monitoring sessions and to higher number of calls per cell. It can also be noticed that the application for measuring the active set size distribution of cells is an excellent opportunity to calculate average active set sizes of single calls as required for calculation of primary/secondary traffic.

The average active set size can be computed for each active set, which means for each particular radio connection. But the aggregation of average active set size at cell level also makes sense to analyse the soft handover situation in the cell. Table 2.22 gives an example for three selected cells of the network.

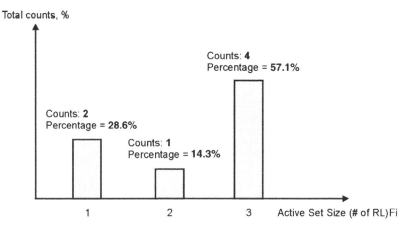

Figure 2.26   Histogram for active set size distribution case study 2

Table 2.22 Average active set size on cell level
(three selected cells)

Cell	Average active set size
Cell 10001	1.8
Cell 10002	1.6
Cell 10003	2.1

The changing active set size of a single call can be correlated to many different radio quality measurements to find out how soft handover situations interfere with the quality of service. Figure 2.27 shows the correlation of the active set size and uplink BLER in a single call. The active set size varies between one and three radio links in the active set. There are some steps in the active set size graph at approximately active set size = 1.6. These steps occur if radio link addition or radio link deletion is monitored in the middle of a sampling period and value 1.6 represents the average active set size within a sampling period of 2 seconds. This effect can be minimised by choosing a shorter sampling period as long as it is not compensated by the graphic plot function of the GUI.

The combined graph analysis shows that UL BLER is especially high (up to 0.375%) if the UE has two or three radio links in the active set for a longer time. Which cells have been involved in such a soft handover situation can be found out by choosing the cell aggregation level (dimension) in the analysis table shown in Figure 2.28.

Indeed the cell with the highest average active set size also has the highest uplink block error rate. The network operator now wants to know if this measurement result can be seen as an exception or do such measurement results occur generally when this cell is a member

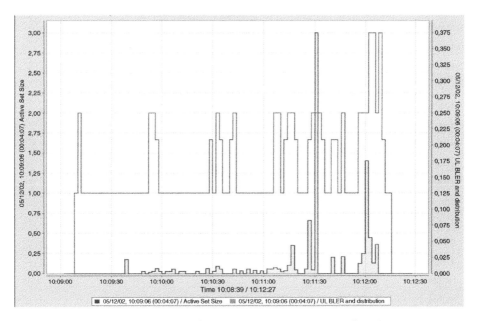

Figure 2.27 UL BLER in % correlated with active set size of call

	Active Set Size	UL BLER ratio [%]
05/12/02, 10:09:06 (00:04:07)	1,5164	0,397 %
Cell-4181	1,1556	0,248 %
Cell-4182	3,0000	2,036 %
Cell-48041	2,1500	0,718 %
Cell-9699	1,5000	0,388 %

Figure 2.28   Active set size and UL BLER per cell involved in a single call

of a UE's active set. The active set size distribution histogram for this particular cell can help to answer this question.

The histogram in Figure 2.29 shows that UEs using this cell always have either two or three radio links in the active set. Hence, all users of this cell are in soft handover. However, the total number of polls have only reached 40: 34 times active set size = 2 and 6 times active set size = 3 has been polled.

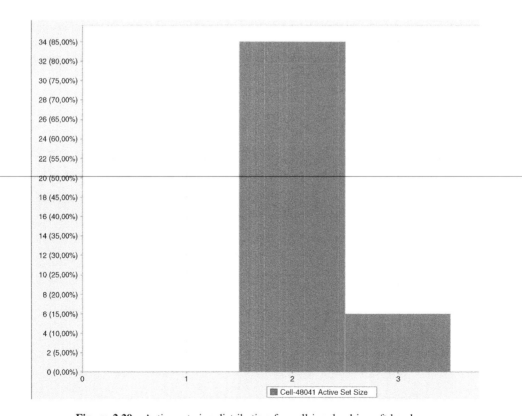

Figure 2.29   Active set size distribution for cell involved in soft handover

The measurement results presented in the histogram indicate that either the observation period has been relatively short and/or that not many UEs have used this cell during the monitoring session. Without doubt a more detailed analysis of the radio environment situation in this cell is necessary, which means the cell must be observed for a longer time and if possible more radio connections using this cell should be monitored. The latter objective can be realised by sending drive test equipment to the desired location, which shows by the way how the benefits of performance measurement software can be combined with drive tests and how this software can help to optimise drive test plans in a very efficient way.

Last but not least it should be mentioned that a further analysis of the overlapping situation in this specific cell can also be done using cell matrices (see section 2.14).

2.7 SOFT HANDOVER SUCCESS AND FAILURE ANALYSIS

If a closer look is taken at what is usually called soft handover it becomes clear that basically there are two different partial procedures belonging to a soft handover. These procedures are called radio link addition and radio link deletion. In the case of radio link addition a new radio link is added to the active set, in the case of radio link deletion a radio link is removed from the current active set of a UE. There is also a special case called radio link substitution (as a rule triggered by reporting RRC measurement event 1C), which means that following a single event-triggered measurement report the worst radio link of an active set is removed while subsequently a better quality radio link is added to the active set of the UE.

Usually RRC protocol messages are seen as events related to these procedures. These messages are RRC Active Set Update Request that indicates a soft handover attempt, RRC Active Set Update Complete that indicates soft handover success and RRC Active Set Update Failure that indicates failed radio link addition. For radio link deletion no dedicated failure event is defined. It is subject of definition if no answer from UE to RRC Active Set Update perviously sent by SRNC must be seen as a failure event. In this case typically the same RRC Active Set Update message is sent again after expiry of a timer on RNC side. The repeated RRC Active Set Update message can be distinguished from the previous one on behalf of its different RRC transaction identifier value.

In RRC Active Set Update Request the message sequences rl-AdditionInformationList and rl-RemovalInformationList are used to determine cases of radio link addition and radio link deletion. The appropriate cell related to these procedures is named by its primary scrambling code as shown in message example 2.13.

Message example 2.13 RRC Active Set Update Request for soft handover radio link deletion

```
| TS 29.331 DCCH-DL (2002-03) (RRC_DCCH_DL) activeSetUpdate (= activeSetUpdate)      |
| dL-DCCH-Message                                                                    |
| 2 message                                                                          |
| 2.1 activeSetUpdate                                                                |
| 2.1.1 r3                                                                            |
| 2.1.1.1 activeSetUpdate-r3                                                          |
| ***b2*** | 2.1.1.1.1 rrc-TransactionIdentifier      | 0                            |
| -1001010 | 2.1.1.1.2 maxAllowedUL-TX-Power          | 24                           |
| 2.1.1.1.3 rl-RemovalInformationList                                                |
| 2.1.1.1.3.1 primaryCPICH-Info                                                      |
| ***b9*** | 2.1.1.1.3.1.1 primaryScramblingCode      | 426                          |
```

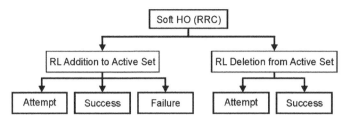

Figure 2.30 Overview of soft handover protocol events

RRC Active Set Update Complete does not contain any specific information element. It is the task of the call trace module to filter out RRC messages belonging to the same connection. In the case of a failed radio link addition the UE will reply with RRC Active Set Update Failure and this message will contain a failure cause value. The occurrence of different failure cause values can be counted using a separate counter for each possible failure cause. Figure 2.30 gives an overview of soft handover analysis as also defined in 3GPP 32.403.

As long as only communication between UE and SRNC is analysed these events are sufficient, but from the above overview two major problems can be identified if there is a need for a more detailed analysis:

1. RRC protocol events do not allow to determine if the performed handover is:
   (a) a softer handover;
   (b) an intra-RNC soft handover; or
   (c) an inter-RNC soft handover involving the Iur interface.
2. If the soft handover procedure has (already) failed before the SRNC sends any RRC message, e.g. due to problems in the NBAP or ALCAP protocol layer, this will not be reflected by the counter defined in the above scheme. In other words: only soft handover failures caused by the UE and radio transmission conditions in cell will be counted, not soft handover failures caused by UTRAN.

Figure 2.31 illustrates the intra-RNC soft handover procedure and highlights the communication that takes place between different protocol entities of the involved network elements including the UE.

Looking at this figure it can be seen that soft handover is not just a single protocol procedure, but rather a complex pattern of different signalling procedures between network elements that interact to perform the handover.

In the first step the UE sends an RRC measurement report indicating that the measured CPICH Ec/N0 value of Cell 2 has entered the reporting range, which means that the RNC is asked to initiate soft handover. While the UE sends this measurement report its radio signal is not only received by Cell 1 but also by Cell 2, however, from the point of view of Cell 2 this signal is still part of the noise/interference in the uplink frequency band, because the cell has not been ordered to decode this UE's signal.

In the second step the RNC starts the soft handover procedure by performing the NBAP radio link setup procedure towards Cell 2. During this procedure Cell 2 is ordered to decode the UE's radio signal identified by the UE's uplink scrambling code. Once the cell has found this uplink channel it sends an NBAP Radio Link Restore Indication message to the RNC.

Soft Handover in Steps

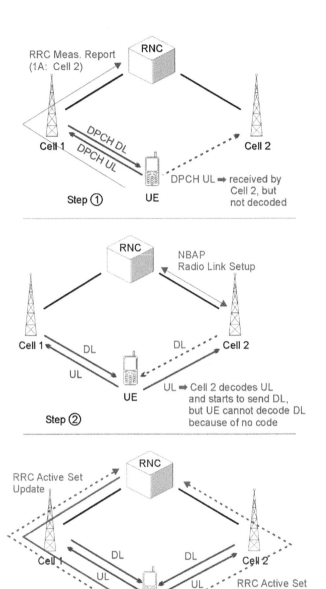

Figure 2.31   Intra-RNC soft handover procedure

In addition, Cell 2 starts to send the downlink dedicated physical channel (DPCH) of the new radio link. However, this DL DPCH cannot be decoded by the UE, because the UE still does not know downlink channelisation code of new radio link.

The downlink channelisation code and primary scrambling code of the new cell, which has already sent the downlink DPCH, are now transmitted to the UE by the RNC using the RRC Active Set Update Request message. After receiving this message the UE adds a new radio link to its active set and starts to decode the DPCH sent by Cell 2. To confirm that radio link addition on the UE side has been successful the UE sends an RRC Active Set Update Complete message, which is usually seen as a soft handover success event by performance measurement software. Due to the special nature of radio transmission in soft handover situation, transport blocks containing portions of the RRC Active Set Update Complete message are transmitted using both uplink dedicated physical (data) channels (DPDCH) simultaneously. By performing macro-diversity combining the RNC will delete duplicated transport blocks and reassemble the single RRC Active Set Update Complete message as sent by the UE. The dotted line used in step 3 of Figure 2.31 visualises this process.

It is important to know that there is no rule that NBAP Radio Link Restore Indication will always be monitored before RRC Active Set Update Complete, because synchronisation processes necessary to send/receive radio signals in a proper manner are running independently from each other on uplink and downlink physical channels. Hence, it is also possible that RRC Active Set Update Complete is received before NBAP Radio Link Restore Indication, but in this case transport blocks that carry RRC messages are only monitored on the Iub interface of Cell 1.

Finally it must be taken into account that similar rules apply to soft handover radio link removal. Here the RRC active set update procedure is executed before NBAP radio link deletion. Hence, as a rule the UE stops to receive the DL physical channel earlier than the cell stops to receive the UL signal of the UE. As a result it can be expected that on Iub of the cell that is already deleted from the UE's active set there are still some uplink transport blocks that can be monitored for which macro-diversity filtering rules apply and that may have an impact on UL BLER measurement results as well as being used to reassemble higher layer messages.

Now it is a question of definition which message is seen as an initial soft handover attempt. The RRC measurement report including event-ID is a good candidate, but there is one limitation. In the case that these reports are sent periodically instead of event-triggered (no event-ID included in message) it is very difficult to identify which measurement report should have triggered the start of the handover procedure that is executed and controlled by the SRNC. To detect a potentially soft handover situation that has been ignored by the SRNC for periodic measurement reports is not impossible, but very hard to do. It requires the follow up of measurement results reported for different neighbour cells and parameter settings that allow performance measurement software to recognise the handover trigger point as internally calculated by SRNC software. Basically the software assumes a soft handover margin of 3 dB. The cell for which soft handover is triggered is identified by the rule that the received pilot of the new cell must be stronger than the received pilot of the best cell in the active set subtracted by the soft handover margin. Certainly parameters such as hysteresis time, which have an decisive impact on soft handover trigger, need to be taken into account and can only be defined by manual configuration of performance measurement

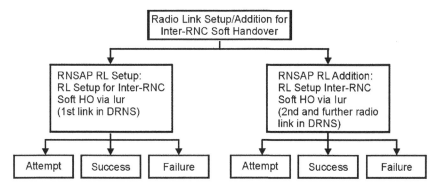

Figure 2.32   Overview of protocol events – inter-RNC soft handover

software. Due to these difficulties it must be questioned if the effort in software development really equals the achieved results.

Once handover trigger based on CPICH measurement is detected the RNC will start the NBAP radio link setup procedure in the case of soft handover. If this NBAP radio link setup procedure has been triggered by the RNSAP Radio Link Setup Request or RNSAP Radio Link Addition Request message, we see an inter-RNC soft handover procedure (as shown in Figure 2.32) and on the RNSAP layer handover may already fail prior to the RRC Active Set Update Request sent by the SRNC to the UE.

If instead of NBAP radio link setup the NBAP radio link addition procedure is monitored, a softer handover is performed. In the case of softer handover there is usually no new AAL2 SVC established as physical transport bearer on Iub, but it is optionally possible. It is not monitored because nearly all networks seem to be optimised successfully in a way that for softer handover diversity combining is activated in Node B by a parameter included in the NBAP Radio Link Addition Request.

All NBAP and RNSAP procedures define unique counters for Attempt (Initiating Message), Success (Successful Outcome) and Failure events (Unsuccessful Outcome). The only procedures that are left to be analysed during soft handover signalling pattern monitored on UTRAN are ALCAP establishment procedures to set up and delete AAL2 SVC, also known as the Iub or Iur physical transport bearer. Here an error in the procedure is indicated if the answer to the AAL2L3 Establish Request message is Release Confirm containing an error cause value that differs from release cause values 'normal, unspecified' and 'normal call clearing' as they are monitored when AAL2 SVC is deleted without exception.

All in all it remains very hard to distinguish if an NBAP radio link setup or ALCAP establishment procedure is executed to prepare soft handover or if they are part of other procedures such as initial DCH set up, channel type switching or hard handover. Indeed, the subsequent RRC active set update procedure is the only reliable indication that soft handover has proceeded.

From the point of view of the author of this book it does not make much sense to define a soft handover success/failure ratio formula that combines all possible protocol procedures monitored on different layers and interfaces. It would rather be advisable to analyse different procedures of different protocols separately and to distinguish – based on subsequent signalling patterns – if e.g. NBAP radio link setup has been performed to prepare soft handover or any other multi-layer procedure. This would result in a definition of a subset of counters for NBAP radio link setup attempt/success/failure events involved in soft handover, another subset of the same event involved in initial DCH set up, another subset for events involved in hard handover etc. Such an implementation also allows a better and faster root cause analysis to troubleshoot the network compared to a combined soft handover failure ratio that does not indicate in which particular network element or protocol entity the failure has occurred. The same implementation strategy can also be used to identify root causes of e.g. radio link failures that often indicate dropped calls and may also be seen as a final result of erroneous soft handover procedures as shown in Figure 1.10.

2.8 INTER-FREQUENCY HARD HANDOVER SUCCESS AND FAILURE RATES

When discussing inter-frequency handover scenarios the most important question is: Which message is used to perform inter-frequency handover? This problem is caused by missing detailed descriptions of inter-frequency handover scenarios in 3GPP standards. 3GPP 25.331 *Radio Resource Control* contains only a very generic description of how to perform hard handovers. It says that basically the following RRC procedures can be used to perform hard handover: radio bearer Setup, radio bearer deletion, radio bearer reconfiguration, transport channel reconfiguration and physical channel reconfiguration. It is the classical problem already highlighted in Figure 1.8 and the authors of 3GPP 32.403 have just copied generic definitions written in 3GPP 25.331. There is no answer to the above question in these documents.

To find the answer an understanding of radio interface procedures is necessary. To perform inter-frequency handover (which is always a hard handover) means that the UE is moving from a source cell working on UTRAN frequency f_1 into a target cell working on UTRAN frequency f_2. To perform this change it is not necessary to set up a new radio bearer, because the radio bearer was already set up when the UE starts to work on frequency f_1. There is also no need to delete the radio bearer, because there is no need to stop transmission of user data via UTRAN, the target is merely to change the frequency used for transmission. Due to these two facts the RRC radio bearer setup and radio bearer deletion procedures can be excluded, because it is very unlikely that they will be used for inter-frequency hard handover. There might be a possible scenario when the radio bearer is deleted in the source cell and a new radio bearer is set up in the target cell, but this would take much more time than a reconfiguration procedure and hence would not be very efficient.

Looking at the radio bearer reconfiguration procedure it becomes clear that the main purpose of this procedure is to change the quality of service attributes assigned to radio links, e.g. to adjust the window size and timer parameters of radio bearers using RLC acknowledged mode. This is again not what we are looking for in the case of inter-frequency handover.

A similar statement is given after looking at transport channel reconfiguration, which is usually used to change the transport format sets of DCHs, e.g. to provide higher data transmission rates on the radio interface or to perform channel type switching, which means changing the mapping of the DTCH onto underlying transport channels.

If it is necessary to change the physical parameters of connections such as used codes without changing transport channel parameters the physical channel reconfiguration procedure is executed. And such a physical parameter is the UMTS absolute radio frequency channel number (uARFCN), which indicates on which uplink and downlink frequency bands FDD radio signals in a cell are transmitted. Cells using the same frequency bands have the same uARFCN values. Knowing this it becomes clear that the handover message we are looking for is most likely an RRC Physical Channel Reconfiguration Request that orders the UE to move to a cell with a different uARFCN. Based on manufacturer-specific implementations of RNC software the previously mentioned radio bearer reconfiguration and transport channel reconfiguration messages can be used in the same manner to perform inter-frequency handover, because physical parameters can also be optionally included in these messages.

Now it is known that it is necessary to look for new uARFCN values included in reconfiguration request messages and the remaining question is: How can it be detected that an included uARFCN value is a different value than the one currently used by the UE for radio transmission? There are two general solutions: either uARFCN changes are tracked for each UE as changes in the active set are tracked. This could be done in the same application, but now more attributes of a single UE need to be stored temporarily. Alternatively it is possible to identify typical multi-layer signalling sequence patterns. These patterns can be recognised by looking at a typical inter-frequency handover scenario that will be explained in the next paragraphs.

Before analysing the handover itself it is necessary to look back at the RRC connection establishment as shown in Figure 2.33. After the UE has sent an RRC Connection Request (1) the network establishes a radio link in the serving cell using the NBAP radio link setup procedure (2) and sends an RRC Connection Setup message (3) to establish signalling

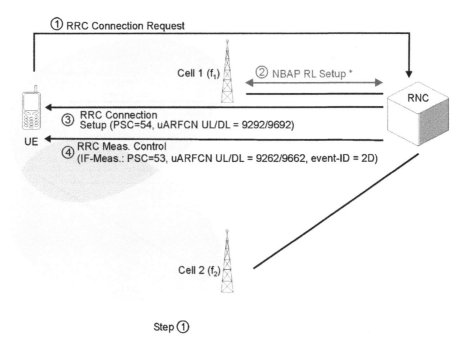

Step ①

Figure 2.33 Inter-frequency handover step 1

radio bearers between the UE and SRNC. This RRC Connection Setup message also contains a cell identifier used on the radio interface: the primary scrambling code (example: PSC = 54). And as an optional parameter uARFCNs for the uplink and downlink frequency band of an FDD cell are also embedded in the same message. In the case of uARFCN tracking these values related to a single connection/UE can now be stored in a background application.

Once RRC connection is successfully established the SRNC assigns measurement tasks to the UE using one or more RRC Measurement Control messages. RRC Measurement Control (4) contains inter-frequency measurement settings for a list of neighbour cells working on a different frequency than the currently used cells of the active set. In the example the active set contains only one cell (PSC = 54). In addition to the neighbour cell list event-IDs expected to be reported to the SRNC are included in a Measurement Control message. Here event-ID 2D is found. This measurement event is explained by 3GPP as 'The estimated quality of the currently used frequency is below a certain threshold'.

What 3GPP specifications do not say is what happens if the currently used frequency is below a certain threshold. For the UE this is a trigger event to start radio quality measurement of the neighbour cells working on a different UTRAN frequency or using a different RAT like GSM or CDMA2000.

This happens indeed after the UE has reported event 2D to SRNC (5) as seen in Figure 2.34. There are two options that allow the UE to monitor cells with a different frequency/RAT: either the UE has a second antenna plus a second RF receiver unit integrated,

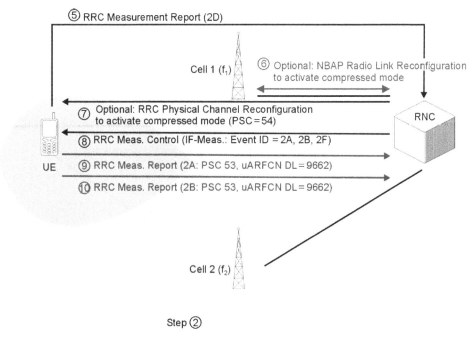

Figure 2.34   Inter-frequency handover step 2

which can be switched on whenever it becomes necessary for inter-frequency or inter-RAT measurements. Or if the UE has only one antenna/RF receiver it is necessary to 'jump' to a different frequency at a certain time, stay there as long as it is necessary to monitor the CPICH and then turn back to the original frequency. This process can be compared to zapping between different channels on a radio or TV. However, due to the fact that the UE simultaneously has an active connection running that is interruption-sensitive (e.g. a voice call), this kind of 'zapping' must happen incredibly fast, within milliseconds. And it must be ensured that during the time the UE is measuring on different frequencies the cell will not continue to send data for ongoing connection on downlink. Hence, to interrupt an ongoing call there needs to be a process synchronised between the UE and the cell. For this reason NBAP radio link reconfiguration (6) and RRC physical channel reconfiguration (7) are optionally executed to activate a so-called compressed mode (which allows 'zapping') and align compressed mode parameters such as gap interval. The gap interval is the time of data transmission interruption due to compressed mode measurements. A new RRC Measurement Control message (8) is sent to the UE by the SRNC, which now orders the mobile to measure on a different frequency identified by uARFCN for uplink and downlink and to report events 2A and 2B if the measured signal level exceeds defined thresholds.

Inter-frequency handover is triggered if the UE sends an RRC measurement report containing event-ID 2A: 'Change of best frequency' (9). This is once again not a change itself, but a request of the UE to change to a currently non-used frequency. In addition, the UE compares the measured threshold of the currently used frequency (f_1) and the non-used frequency (f_2) and if the level of f_2 is above the level of f_1 another RRC measurement report is sent containing event-ID 2B (10), which means: 'The estimated quality of the currently used frequency is below a certain threshold *and* the estimated quality of a non-used frequency is above a certain threshold'. Using a popular explanation it could be said that this second reported event is the UE's emphatic request to perform inter-frequency handover. It is still unknown if the RNC would also react to a single measurement report containing event 2A or 2B. This seems to be a manufacturer-specific implementation in RNC software. Indeed, the identity of the cell working on a different frequency (in example: primary scrambling code [PSC] = 53) and the used downlink frequency of cell 2 (uARFCN DL = 9662) are included in each of the RRC measurement reports. This information is crucial for the SRNC to perform inter-frequency handover.

Before the handover of the UE can be executed it is necessary to prepare the target cell of the handover procedure to send/receive radio signals using specific codes of this connection as shown in Figure 2.35. Cell 2 gets the necessary code information and other important parameters on behalf of an NBAP radio link setup procedure (11). This NBAP procedure also triggers the subsequent establishment and synchronisation of the Iub physical transport bearer (AAL2 SVC), which is not shown in the figure.

To order UE to change the cell as well as the used frequency the SRNC sends an RRC Physical Channel Reconfiguration Request message (12) that once again contains the primary scrambling code of cell 2 and the uARFCN for the uplink and downlink frequency band of this new cell. When this message is received by the UE it switches to the new cell/frequency and sends an RRC Physical Channel Reconfiguration Complete message (13), and is already using the new cell/frequency to the SRNC. This message completes the handover procedure. A failure in the handover procedure on the UE side is indicated by an RRC

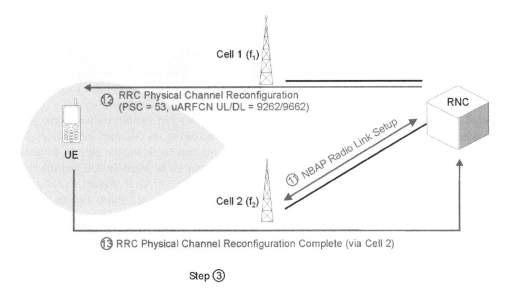

Figure 2.35   Inter-frequency handover step 3

Physical Channel Reconfiguration Failure message. However, it may also happen that the UE switches to the new frequency, but for some reason it is not able to establish connection with cell 2 and (e.g. due to mobility) it is also not able to change back to the old radio link of cell 1. The UE disappears, so to speak. In this case an NBAP Radio Link Failure Indication sent by Node B of cell 1 would be monitored while NBAP Radio Link Restore Indication related to established radio link of cell 2 is missed as well as RRC Physical Channel Reconfiguration Complete sent via cell 2. The problem is that in the case of a successful handover the radio link in cell 1 needs to be deleted, but due to the reaction time of the RNC there is often also a NBAP Radio Link Failure Indication from cell 1 monitored in the case of a successful handover – just because the RNC cannot send the order to delete the old radio link fast enough. Therefore cell 1, which has never received any information about the planned or executed handover, tells the RNC that it no longer receives an uplink signal of UE and hence from the point of view of cell 1 UE is lost.

Based on these three messages the following protocol events for success/failure formulas can be defined:

- *IFHO_Attempt*: RRC Physical Channel Reconfiguration Request triggered by RRC measurement report events 2A and 2B.
- *IFHO_Success*: RRC Physical Channel Reconfiguration Complete message bound to previous IFHO_Attempt by the same RRC transaction identifier or identified by the call trace filter as belonging to the same UE connection.
- *IFHO_Failure_UE*: RRC Physical Channel Reconfiguration Failure message bound to the previous IFHO_Attempt by the same RRC transaction identifier or identified by the call trace filter as belonging to the same UE connection.
- *IFHO_RL_Failure:* NBAP Radio Link Failure Indication following IFHO_Attempt without monitoring IFHO_Success event.

The formulas are defined as follows:

$$\text{IFHO Success Rate} = \frac{\sum \text{IFHO Success}}{\sum \text{IFHO Attempt}} \times 100\% \qquad (2.15)$$

$$\text{IFHO Failure Rate} = \frac{\sum (\text{IFHO Failure UE} + \text{IFHO RL Failure})}{\sum \text{IFHO Attempt}} \times 100\% \qquad (2.16)$$

However, as easy as it is to define and write a formula, it is as difficult to understand and implement the check of a signalling pattern indicating Attempt and Failure events. Another problem might come up when discussing the aggregation level of these KPIs. It is useful to analyse it per call, this is clear, and there are three options: since it is an RRC protocol procedure and RRC messages can be exchanged in soft handover situations it is possible that there is not just a single source cell, but a source active set. Due to this, Attempt event could be shown related to the best cell of this active set, but the best cell is changed during handover and hence, Success event would be displayed to a new best cell that is the target cell. If Attempt counters are counted in source cell(s) while Success counters are counted in the target cell it might become difficult for performance measurement software to compute Success ratio per cell. One possible solution is to aggregate protocol event counters not at the cell at which they have been monitored, but at a cell defined by identifiers found in handover procedure messages: either show all events related to the source cell or related to the target cell. A variant to showing counters as well as ratios related to the target cell is probably the most meaningful, because if handover is not successful it is much more often due to radio conditions in the target cell than due to errors in UE software (that occur according to load/stress test results in fewer than 1 out of 10 000 cases). Finally, there is another option that displays both source and target cell of the handover procedure. This option is the handover cell matrix as described at the end of section 2.14.

2.9 CORE NETWORK HARD HANDOVER SUCCESS AND FAILURE RATES

The handover procedures discussed so far in the previous chapters have been always executed within the UTRAN, which means that UEs have never left the area controlled by a single core network element such as the MSC or SGSN and Iub and Iur interfaces have been involved in handover signalling procedures. Now emphasis is set on the analysis of handover procedures switching in the core network. The following scenarios will be discussed:

3. 3G-3G Intra-frequency hard handover/relocation
4. 3G-2G Inter-RAT Handover for CS and PS services
5. 2G-3G Inter-RAT Handover for CS and PS services

2G in this context stands as a synonym for GSM, inter-RAT handover to/from CDMA2000 is beyond the scope of this book.

2.9.1 INTRA-MSC AND INTER-MSC HARD HANDOVER (3G-3G)

Intra-frequency hard handover becomes necessary if inter-RNC handover needs to be performed, but there is no Iur interface available between the source and target RNC. In such cases (no Iur) each inter-RNC intra-frequency handover must inevitably be a hard handover although the target cell is working on the same frequency as the source cell. If there is no Iur interface the target RNC cannot act as a drift RNC and lend its radio resources to the SRNC. Hence, the target RNC needs to become the SRNC immediately.

Inter-RNC intra-frequency hard handovers are triggered by the same measurement report event-IDs as softer and soft handover radio link additions: 1A. In addition (also true for softer and soft handover) event-ID 1E needs to be considered as another possible trigger if the new strong cell belongs to the detected set and has not been part of the neighbour cell information list sent to the UE by the SRNC.

For performance measurement software it is a problem that it only monitors if an RNC receives an RRC measurement report containing event-ID 1A for a cell identified by primary scrambling code = 52. Due to the fact that there is a maximum of only 512 different primary scrambling codes it is possible that PSC = 52 could be used e.g. five times for cells controlled by the same RNC. One of these five cells is the target cell of the handover procedure, but as long as there is no NBAP Radio Link Setup Request for this cell monitored the performance measurement software can only guess identity of this cell. So how does the RNC know the unique identity of the target cell if there are five options?

For each cell controlled by the RNC a neighbour cell list needs to be manually configured. This neighbour cell list not only contains the primary scrambling codes as monitored on the Iub interface in RRC measurement messages, but also detailed information of each cell that allows message routing to the target RNC in the case of inter-MSC handover. The full neighbour parameter list can only be monitored on the Iur interface during the RNSAP radio link setup procedure as shown in message example 2.14.

Message example 2.14 RNSAP radio link setup response including neighbour cell info list

ID Name	Comment or Value	
UMTS RNSAP acc. R99 TS 25.423 ver. 3.7.0 (RNSAP370) successfulOutcome		
rnsapPDU		
1 **successfulOutcome**		
1.1 procedureID		
1.1.1 procedureCode	**id-radioLinkSetup**	
1.4.1.4.3.1.3.14 neighbouring-UMTS-CellInformation		
1.4.1.4.3.1.3.14.1 sequenceOf		
1.4.1.4.3.1.3.14.1.1 id	id-Neighbouring-UMTS-CellInformat...	
1.4.1.4.3.1.3.14.1.2 criticality	ignore	
1.4.1.4.3.1.3.14.1.3 value		
1.4.1.4.3.1.3.14.1.3.1 rNC-ID	124	
1.4.1.4.3.1.3.14.1.3.2 neighbouring-FDD-CellInformation		
1.4.1.4.3.1.3.14.1.3.2.1 neighbouring-FDD-CellInformationItem		
1.4.1.4.3.1.3.14.1.3.2.1.**1 c-ID**	**15009**	
1.4.1.4.3.1.3.14.1.3.2.1.2 uARFCNforNu	9762	

Message example 2.14 (*Continued*)

ID Name	Comment or Value	
1.4.1.4.3.1.3.14.1.3.2.1.3 uARFCNforNd	10712	
1.4.1.4.3.1.3.14.1.3.2.1.4 **primaryScramblin..**	**77**	
1.4.1.4.3.1.3.14.1.3.2.1.5 primaryCPICH-Power	314	
1.4.1.4.3.1.3.14.1.3.2.1.6 cellIndividualOf..	0	
1.4.1.4.3.1.3.14.1.3.2.1.7 txDiversityIndic..	false	
1.4.1.4.3.1.3.14.1.3.2.2 neighbouring-FDD-CellInformationItem		
1.4.1.4.3.1.3.14.1.3.2.1.1 **c-ID**	**15006**	
1.4.1.4.3.1.3.14.1.3.2.1.2 uARFCNforNu	9762	
1.4.1.4.3.1.3.14.1.3.2.1.3 uARFCNforNd	10712	
1.4.1.4.3.1.3.14.1.3.2.1.4 **primaryScramblin..**	**96**	
1.4.1.4.3.1.3.14.1.3.2.!.5 primaryCPICH-Power	316	
1.4.1.4.3.1.3.14.1.3.2.1.6 cellIndividualOf..	0	
1.4.1.4.3.1.3.14.1.3.2.!.7 txDiversityIndic..	false	
1.4.1.4.3.1.3.14.1.3.2.3 neighbouring-FDD-CellInformationItem		
1.4.1.4.3.1.3.14.1.3.2.3.1 **c-ID**	**14259**	
1.4.1.4.3.1.3.14.1.3.2.3.2 uARFCNforNu	9762	
1.4.1.4.3.1.3.14.1.3.2.3.3 uARFCNforNd	10712	
1.4.1.4.3.1.3.14.1.3.2.3.4 **primaryScramblin..**	**100**	
1.4.1.4.3.1.3.14.1.3.2.3.5 primaryCPICH-Power	317	
1.4.1.4.3.1.3.14.1.3.2.3.6 cellIndividualOf..	0	
1.4.1.4.3.1.3.14.1.3.2.3.7 txDiversityIndic..	false	

As can be seen in the message example there are three different cells controlled by RNC 124 listed using their c-ID (as used in NBAP and RNSAP signalling) plus their primary scrambling codes. Having RNC-ID and c-ID each cell is identified uniquely. Appropriate routing information for transmission of handover request messages is easy to derive from RNC-internal routing tables for handover/relocation scenarios when the RRC measurement report containing only PSC value is received.

The most common variant of intra-frequency hard handover is 3G-3G inter-MSC handover triggered by event-ID 1A if it is detected that the target cell reported together with event-ID is controlled by an RNC connected to a different MSC than the current SRNC. There is no chance of performing a radio link addition, because usually in this case there is no Iur interface between the SRNC (source RNC) and the target RNC. Instead the SRNC decides – based on information found in the cell neighbour list and routing tables – to start an RANAP relocation/handover procedure.

A relocation is a special type of handover that is characterised by changing the serving cell of the radio connection plus re-routing of bearers for signalling and user data transport to the new serving cell.

A relocation/handover procedure is executed in four major steps (no matter how many signalling messages are involved in each step in detail). Figure 2.36 shows these steps for intra-MSC hard handover/relocation for reasons of graphical simplicity. If another MSC

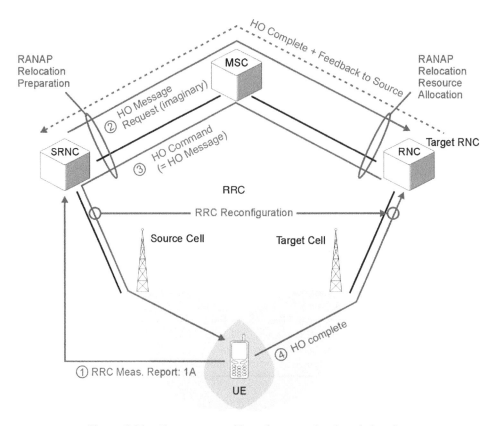

Figure 2.36   Abstract steps of intra-frequency handover/relocation

(inter-MSC handover) is involved this second MSC is located between the shown MSC and the target RNC and an imaginary handover request as well as a handover command are transparently routed through this additional network element. The relocation of transport bearers is only performed on Iu and Iub interfaces. The originally serving MSC of the call remains the anchor MSC of the call and continues to control the call link towards the gateway MSC (GMSC) after successful handover/relocation.

The procedure starts with an RRC measurement report (1). Following this the SRNC sends a handover message request to the target RNC (2). This handover message request is of 'imaginary' nature, because it is not directly visible in the RANAP signalling messages. Two different IuCS interfaces and two different RANAP procedures are involved. On the old IuCS the handover message request is 'hidden' in the Relocation Required message, on the target IuCS the handover request is forwarded using the RANAP Relocation Request.

Note: 3GPP standard documents use message names in procedure descriptions, but these messages are encoded using ASN.1. Looking at ASN.1 only four different RANAP message codes exist: Initiating Message, Successful Outcome, Outcome and Unsuccessful Outcome. These message codes are combined with different procedure codes. Each message name used *in 3GPP specifications has its unique ASN.1 message code/procedure code combination. Hence, **RANAP Relocation Required** message is encoded as **RANAP InitiatingMessage (ProcedureCode: RelocationPreparation)**. This principle is also valid for other UTRAN protocols. In message example 2.14, for instance, RNSAP Radio Link Setup Response message is presented, which is encoded as RNSAP SuccessfulOutcome (ProcedureCode: Radio Link Setup).*

From the 'imaginary' request coming from the SRNC the target RNC constructs a handover command message to be sent to the UE. However, the target RNC does not have a radio connection with the UE – but the SRNC has! Following this, the handover command (which is the previously requested handover message) is sent via the MSC, SRNC and source cell to the UE, because this is the only available radio connection (3).

Typically for Intra-frequency hard handover this handover command message is an RRC Physical Channel Reconfiguration Request containing new codes to be used for the radio link in the target cell and additional new identifiers such as a new u-RNTI, which will change due to the fact that the SRNC of connection will change.

After the UE has received the hard handover command message (e.g. RRC Physical Channel Reconfiguration Request) it starts to send/receive data using configuration parameters received in this message. For successful reconfiguration the UE sends the handover complete message (e.g. RRC Physical Channel Reconfiguration Complete message) via the new cell/new Iub to the target RNC. At the target RNC, which has now become the new SRNC of the connection, the loop between the handover command and handover complete is closed and from the point of view of the radio connection handover was successful. There is just one more thing to do: the transport resources on the previously used interfaces (old IuCS, old Iub) need to be deleted as well as all call-specific information in RNC and MSC protocol entities. This information to be deleted includes all the parameters of the RRC context of the UE and assigned (and now successfully relocated) RABs.

To delete the RRC context and RABs in the old SRNC the feedback of successfully completed handover (4) is delivered by the target RNC (new SRNC) to the core network element (MSC) and from core network back to source RNC (old SRNC). Once again RANAP messages are used to forward this information and subsequently radio links in the old cell are deleted as well as RABs.

Regarding the processing time of NBAP radio link deletion the same problem as already described in for inter-frequency handovers is true: if it takes too long until the old SRNC starts NBAP radio link deletion triggered by feedback about successful relocation the source cell will complain by sending NBAP Radio Link Failure Indication that contact with the UE has been lost. On the other hand the same risk exists that the UE is really lost due to the handover attempt and the handover complete message is not monitored on the new Iub.

The RAB is released and performs the RANAP Iu-Release procedure. An appropriate release cause used in this procedure following a successful handover is 'successful relocation'. In the past it was observed that some NEM implemented cause value 'normal release', but this is not compliant with 3GPP standards, because 'normal release' belongs to the group of NAS causes that should only be used if an NAS protocol event (e.g. Call

Control DISCONNECT) is the reason for terminating an active connection. A relocation/handover is a radio network related procedure, because it is performed due to the mobility of the UE. Following this the radio network layer cause 'successful relocation' must be used in an appropriate Iu-Release procedure.

All details of the procedure are now known and protocol event counters and KPI formulas can be defined, but there are different approaches. On the one hand it is possible to define easy success rates for the procedure using the following protocol events:

- *InFHHO_Attempt*: RANAP Relocation Required containing sourceRNC-to-targetRNC-transparent-container. This container is used to transmit necessary RRC context parameters from source to target RNC and as explained in the next section a similar container for inter-RAT handover looks completely different. Hence, based on the existence of this transparent container 3G-3G handover/relocation can be identified easily.

  Note: The same sourceRNC-to-targetRNC-transparent-container can also be used for 3G_Intra-/Inter-MSC inter-frequency handover – if the target cell is working on a different UTRAN frequency than the source cell. If such a handover is possible in the network (depends on network architecture) separate counters for intra- and inter-frequency handover switching in the core network can be defined by checking the event-IDs in the RRC measurement report that triggered the sending of RANAP Relocation Required or are based on the check of uARFCN values provided by the active set tracking the background application (as discussed in the section about inter-frequency handover).

- *InFHHO_Success*: an overall success rate KPI RANAP Iu-Release Command including cause value 'successful relocation' can be seen as a good success event, because on the one hand following the reception of this message the RNC will delete all resources related to included RAB-IDs and on the other hand monitoring this message indicates that hard handover/relocation to target cell/RNC has been successful without any doubt. In the (unlikely) case that the SRNC has some problems relaeasing the old RABs and related radio resources this will not have any impact on the active connection between the UE and the network.

- ~~*InFHHO_Failure_UE*: any RRC Reconfiguration Failure message sent as response to~~ RRC Reconfiguration Request (used as handover command message) must be seen as a failure indication from the UE side.

- *InFHHO_RL_Failure*: if the UE loses contact with the network after it has received the handover command an NBAP Radio Link Failure Indication is expected to be monitored while *InFHHO_Success* event is missed. Due to the possibility that an NBAP Radio Link Failure Indication may also be sent in the case of successful handover it is necessary to perform a careful check to see if this message indicates a failure event or not. If NBAP Radio Link Failure Indication is monitored for a UE that has already received a handover command (Attempt) it needs to be checked if RAB(s) related to this UE have already been ordered to be deleted by the MSC using RANAP Iu-Release (cause = 'successful relocation'). If RANAP Iu-Release has not been monitored a timer needs to be started to check if the success event is monitored within a certain time frame. Otherwise handover is considered to have failed and subsequently it needs to be considered to drop the call, too.

Formulas are easily defined:

$$\text{InFHHO Success Rate} = \frac{\sum InFHHO\ Success}{\sum InFHHO\ Attempt} \times 100\% \qquad (2.17)$$

$$\text{InFHHO Failure Rate} = \frac{\sum(InFHHO\ Failure\ UE + InFHHO\ RL\ Failure)}{\sum InFHHO\ Attempt} \times 100\% \quad (2.18)$$

2.9.2 3G-2G INTER-RAT HANDOVER FOR CS AND PS SERVICES

If the UE is moving and there is no other UTRAN cell offering sufficient quality for handover of an intra-frequency or inter-frequency type a neighbour GSM cell might be the best solution for handover. Basically inter-RAT relocation/handover is the same procedure as explained before with some minor, but decisive differences as illustrated in Figure 2.37.

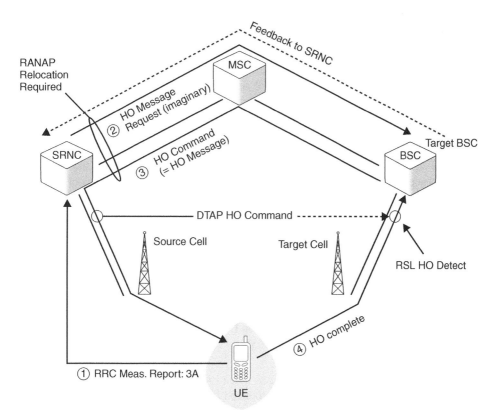

Figure 2.37 Abstract steps of CS inter-RAT handover 3G-2G

2.9.2.1 CS 3G-2G Inter-RAT Handover

The measurement report event-ID that triggers inter-RAT handover from UTRAN to GSM is 3A: 'The estimated quality of the currently used UTRAN frequency is below a certain threshold **and** the estimated quality of the other system is above a certain threshold'.

Handover message request (2) is once again seen as an RANAP Relocation Preparation message on IuCS, but this time a different container is embedded. Instead of the sourceRNC-to-targetRNC-transparent-container we find a protocol sequence named oldBSS-to-newBSS-information. This sequence contains handover-relevant information encoded in BSSMAP format and is necessary to construct the BSSMAP Handover Request sent by the MSC to the target BSC.

In this scenario the target BSC has substituted the target RNC. The interface between the MSC and target BSC is the A-interface, the interface between the target cell and the target BSC is Abis.

After handover message request (2) is received in the target BSC this network element will construct a DTAP Handover Command message (3) that is sent back to the SRNC and forwarded by the SRNC to the UE embedded in the RRC Handover from UTRAN to the GSM command. On IuCS interfaces the same DTAP message was previously transported embedded in TargetRNC-to-SourceRNC-Transparent-Container. In other words, if the source RNC sends handover information to the target BSC it uses BSSMAP, the way of expression on the BSC/A-interface, while on the other hand if the target BSC sends handover-related information to the source RNC it uses RANAP, the RNC/IuCS way of expression. Due to this a prerequisite of such a kind of handover is a software update on the BSC side as well, because standard 2G BSC software does not offer such functionality.

Whether the UE switched to the GSM cell can be detected on the GSM Abis interface. The radio signalling link (RSL) protocol will send a Handover Detect message to the BSC that indicates that the cell has been able to establish radio contact with the UE. This RSL Handover Detect is the handover complete message (4). Similar to 3G-3G hard handover scenarios described in the previous section the reception of this message in the BSC triggers a feedback mechanism and the reception of this feedback information on the MSC will trigger the RANAP Iu-Release procedure (cause = 'successful relocation') on the IuCS interface and subsequently the release of all radio and transport network resources of this particular connection in UTRAN.

Protocol events:

- *CS-iRAT-HO_Attempt*: RANAP Relocation Required containing oldBSS-to-newBSS-Information.
- *CS-iRAT-HO_Success:* RANAP Iu-Release (cause = 'successful relocation') monitored for the same RANAP connection as *CS-iRAT-HO_Attempt*. RANAP messages monitored on a single Iu interface belonging to the same connection are bound by a unique pair of SCCP Source Local Reference & Destination Local Reference numbers (see Iu signalling scenarios in Kreher and Ruedebusch, 2005).

There must be three different failure cases taken into account for failure events. For this reason three different failure events are introduced in this section.

CS-iRAT-HO_Failure_1: RANAP Relocation Cancel message. If the traffic channel cannot be established on all involved interfaces of the GSM/EDGE radio access network

(GERAN) this will be indicated by Relocation Preparation Failure (ASN.1 encoded: RANAP UnsuccessfulOutcome, Procedure Code = 'Relocation Preparation'). For root cause analysis it is recommended to check the probably embedded *Inter-System Information Transparent Container* or cause values such as 'no radio resource available in target cell'. After receiving Relocation Preparation Failure it is always up to the SRNC to decide if the relocation procedure will be cancelled. If the SRNC decides to cancel, the RANAP Relocation Cancel message is monitored.

If RANAP Relocation Preparation Failure is received by the SRNC in some cases (e.g. cause value = 'semantic error') another Relocation Required message will be sent immediately to the MSC. If this attempt fails again another Relocation Required is sent and so on. There is an indefinite number of CS-iRAT-HO_Attempt for a single connection in such cases. It must be defined that a failed relocation preparation procedure is only detected if in the end no relocation preparation procedure has been finished successfully. Also, if Successful Outcome of Relocation Preparation is seen after 30 or 40 unsuccessful attempts and it has taken more than 30 seconds to finally get Successful Outcome the relocation procedure must be counted as successful. If possible a separate root cause analysis can be done as follows: compare the target ID information element of Relocation Required messages for which SuccessfulOutcome was monitored with the target ID of Relocation Required messages for which UnsuccessfulOutcome has been seen. Using this method we can identify target cells (or their BSCs) that need to be trouble-shooted (possible root cause: their software is not able to interact with required RANAP protocol versions). Figure 2.38 shows a typical message example that can be used for this kind of trouble-shooting. The target ID highlighted in the message parameters identifies the target cell uniquely. The trace sequence in the upper part of the figure (Short View pane) shows that the target system cannot react to the required relocation because of 'semantic error'.

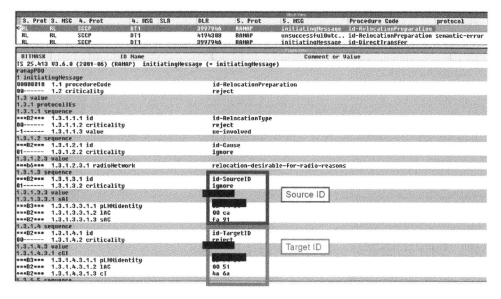

Figure 2.38 RANAP relocation preparation failure trace

CS-iRAT-HO_Failure_2: the second failure event is the RRC Handover from UTRAN Failure message that can be measured on Iub if the RRC handover procedure fails.

CS-iRAT-HO_Failure_3: the third possibility of a failure indication is once again the case that the UE loses contact with the radio network during the handover procedure and we see NBAP Radio Link Failure without having seen a CS-iRAT-HO_SUCCESS event or Handover Complete message on the GERAN Abis interface (if the performance measurement unit is able to monitor GERAN interfaces and call trace application features combined with UTRAN/GERAN monitoring).

Formulas:

$$\text{CS-IRAT-HO Success Rate} = \frac{\sum \textit{CS-IRAT-HO Success}}{\sum \textit{CS-IRAT-HO Attempt}} \times 100\% \tag{2.19}$$

$$\text{CS-IRAT-HO Failure Rate} = \frac{\sum_{x-1}^{x-3} \textit{CS-IRAT-HO Failure x}}{\sum \textit{CS-IRAT-HO Attempt}} \times 100\% \tag{2.20}$$

2.9.2.2 PS 3G-2G Inter-RAT Handover

The main difference between CS inter-RAT handover and PS inter-RAT handover from UTRAN to GSM is that for the PS handover version *no RANAP Relocation procedure is executed* as shown in Figure 2.39.

The handover trigger event is again RRC event-ID 3A (after a number of 2D events have been reported, which have triggered the activation of inter-RAT measurement on RRC level). Instead of starting a relocation the SRNC immediately sends an RRC CellChangeOrderfromUTRAN message to the UE once a neighbour GSM cell has been measured to offer sufficient radio quality for taking over the active connection.

The reason why there is no relocation performed is that there is no dedicated traffic channel for IP payload on GERAN interfaces Gb and Abis. Instead so-called temporary block flows are monitored on Abis whenever IP payload needs to be transmitted between the UE and the network.

Protocol events:

- *PS-iRAT-HO_Attempt*: RRC CellChangeOrderfromUTRAN message.
- *PS-iRAT-HO_Success:* RANAP Iu-Release (cause = 'release-due-to-utran-generated-reason').

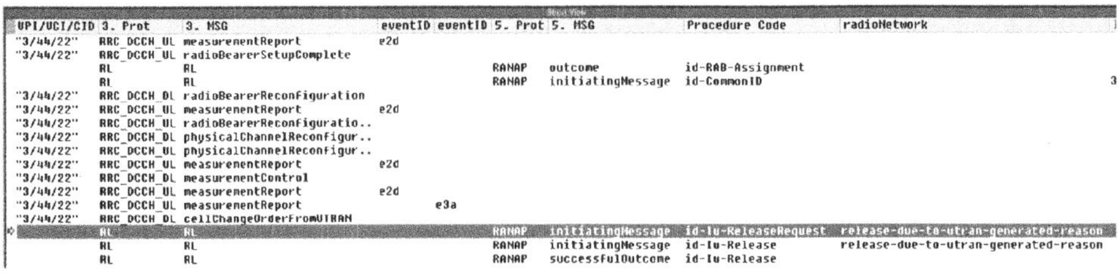

UPI/UCI/CID	3. Prot	3. MSG	eventID	eventID	5. Prot	5. MSG	Procedure Code	radioNetwork	
"3/44/22"	RRC_DCCH_UL	measurementReport	e2d						
"3/44/22"	RRC_DCCH_UL	radioBearerSetupComplete							
	RL	RL			RANAP	outcome	id-RAB-Assignment		
	RL	RL			RANAP	initiatingMessage	id-CommonID		3
"3/44/22"	RRC_DCCH_DL	radioBearerReconfiguration							
"3/44/22"	RRC_DCCH_UL	measurementReport	e2d						
"3/44/22"	RRC_DCCH_UL	radioBearerReconfiguratio..							
"3/44/22"	RRC_DCCH_DL	physicalChannelReconfigur..							
"3/44/22"	RRC_DCCH_UL	physicalChannelReconfigur..							
"3/44/22"	RRC_DCCH_UL	measurementReport	e2d						
"3/44/22"	RRC_DCCH_DL	measurementControl							
"3/44/22"	RRC_DCCH_UL	measurementReport	e2d						
"3/44/22"	RRC_DCCH_UL	measurementReport		e3a					
"3/44/22"	RRC_DCCH_DL	cellChangeOrderFromUTRAN							
	RL	RL			RANAP	initiatingMessage	id-Iu-ReleaseRequest	release-due-to-utran-generated-reason	
	RL	RL			RANAP	initiatingMessage	id-Iu-Release	release-due-to-utran-generated-reason	
	RL	RL			RANAP	successfulOutcome	id-Iu-Release		

Figure 2.39 PS inter-RAT handover 3G-2G

- *PS-iRAT-HO_Failure:* if handover fails the RRC Cell Change Orderfrom UTRAN Failure message is monitored or the already well-known scenario of NBAP Radio Link Failure Indication without having seen Success event.

Formulas for Success and Failure Ratio can be constructed in the same way as demonstrated in inter-frequency handover analysis.

2.10 STATE TRANSITIONS AND CHANNEL TYPE SWITCHING

When analysing state transitions it must be distinguished between protocol states and virtual states defined in performance measurement software. As already mentioned in the section about performance measurement software architecture in Chapter 1 some measurement equipment manufacturers have based a large part of their call analysis on state machines. States used by such software are mostly virtual ones, which means that they have not been defined in any international standard document published by 3GPP or other standards organisations. Such virtual states are beyond the scope of this section with one single exception: the *HSDPA Active* state.

Due to the fact that there is no *HSDPA Active* state defined in 3GPP a virtual state needs to be defined if it is to be analysed when, how often and how long the UE uses HS-DSCH for downlink data transmission.

During call establishment the SRNC sends first RRC state information to the UE embedded in the RRC Connection Setup message. Either the UE is ordered to enter CELL_DCH state and use dedicated channels (with parameters transmitted in the same RRC Connection Setup message) for further signalling exchange with the network, or the UE is ordered to enter CELL_FACH state and use common transport channels RACH and FACH for RRC/ NAS signalling transmission. Which option is used depends on the RNC configuration settings and optimisation targets defined for UTRAN.

If further exchanged signalling leads to a radio bearer setup (in currently installed network configurations) this always requires the UE entering CELL_DCH state after request by the SRNC transmitted to the UE using the RRC Radio Bearer Setup Request message shown in message example 2.15.

While speech calls remain in CELL_DCH as long as a call is active the radio bearers of PS data calls are adapted to currently required data transmission rates whenever the SRNC decides that this is necessary. Procedures defined for such dynamical reconfigurations are all RRC reconfiguration procedures: RRC physical channel reconfiguration, RRC transport

Message example 2.15 RRC radio bearer setup request (RRC State = CELL_DCH)

ID Name	Comment or Value	
TS 29.331 DCCH-DL (2002-09) (RRC_DCCH_DL) radioBearerSetup (= radioBearerSetup)		
dL-DCCH-Message		
2 message		
2.1 **radioBearerSetup**		
2.1.1 r3		
2.1.1.1 radioBearerSetup-r3		
2.1.1.1.1 rrc-TransactionIdentifier	0	
2.1.1.1.2 **rrc-StateIndicator**	**cell-DCH**	

channel reconfiguration and RRC radio bearer reconfiguration. In all these procedures we can monitor RRC state indicators as well as optionally channel mapping options. Sometimes only the spreading factor and transport format sets of DCH are adapted to current needs – as shown in Figure 1.36. Sometimes the type of transport channel which dedicated traffic channels (DTCH) are mapped onto is changed: if IP data volume to be transmitted is expected to be extremely small DTCH is mapped onto common transport channels RACH and FACH. Whenever this happens RRC state is also changed from CELL_DCH to CELL_FACH. Usually the RRC Physical Channel Reconfiguration Request message is used to transmit this state change information, but also the RRC Radio Bearer Reconfiguration Request can be used as shown in message example 2.16.

Message example 2.16 RRC radio bearer reconfiguration (RRC state = CELL_FACH)

ID Name	Comment or Value	
TS 29.331 DCCH-DL (2002-09) (RRC_DCCH_DL) radioBearerReconfiguration		
dL-DCCH-Message		
2 message		
2.1 radioBearerReconfiguration		
2.1.1 r3		
2.1.1.1 radioBearerReconfiguration-r3		
2.1.1.1.1 rrc-TransactionIdentifier	1	
2.1.1.1.2 activationTime	80	
2.1.1.1.3 new-C-RNTI	'0000000000000001'B	
2.1.1.1.4 **rrc-StateIndicator**	**cell-FACH**	
2.1.1.1.5 **rb-InformationReconfigList**		
2.1.1.1.5.1 rB-InformationReconfig		
2.1.1.1.5.1.1 rb-Identity	1	
2.1.1.1.5.1.2 rlc-Info		
2.1.1.1.5.1.2.1 ul-RLC-Mode		
2.1.1.1.5.1.2.1.1 ul-UM-RLC-Mode		
2.1.1.1.5.1.2.2 dl-RLC-Mode		
2.1.1.1.5.1.2.2.1 dl-UM-RLC-Mode	0	
2.1.1.1.5.1.3 rb-MappingInfo		
2.1.1.1.5.1.3.1 rB-MappingOption		
2.1.1.1.5.1.3.1.1 ul-LogicalChannelMappings		
2.1.1.1.5.1.3.1.1.1 oneLogicalChannel		
2.1.1.1.5.1.3.1.1.1.1 **ul-TransportChannelType**		
2.1.1.1.5.1.3.1.1.1.1.1 **rach**	0	
2.1.1.1.5.1.3.1.1.1.2 logicalChannelIdentity	1	
2.1.1.1.5.1.3.1.1.1.3 rlc-SizeList		
2.1.1.1.5.1.3.1.1.1.3.1 explicitList		
2.1.1.1.5.1.3.1.1.1.3.1.1 rLC-SizeInfo		
2.1.1.1.5.1.3.1.1.1.3.1.1.1 rlc-SizeIndex	1	
2.1.1.1.5.1.3.1.1.1.4 mac-LogicalChannelPri..	1	
2.1.1.1.5.1.3.1.2 dl-LogicalChannelMappingList		
2.1.1.1.5.1.3.1.2.1 dL-LogicalChannelMapping		
2.1.1.1.5.1.3.1.2.1.1 **dl-TransportChannelType**		
2.1.1.1.5.1.3.1.2.1.1.1 **fach**	0	
2.1.1.1.5.1.3.1.2.1.2 logicalChannelIdentity	1	

In this message example channel mapping options are also highlighted that are used for channel type switching from DCH to RACH/FACH. In a similar way channel type switching of DL transport channel to HS-DSCH can be performed. It happens whenever the downlink transport channel type is HS-DSCH (instead of FACH in the message example above). However, a prerequisite of using HS-DSCH is that the UE is in CELL_DCH state. Knowing these facts the virtual state *HSDPA Active* can be defined as follows:

HSDPA Active: UE in CELL_DCH and active downlink transport channel = HS-DSCH.

So far it does not seem to be difficult to follow HSDPA activations/deactivations. HSDPA is deactive if the radio bearer/DTCH is mapped onto the downlink dedicated transport channel (DCH) or Forward Access Channel (FACH) again. However, the problem is that radio bearer mapping options are often only transmitted once when the radio bearer is set up or reconfigured for the first time. Then these mapping options are stored in the UE as well as in the SRNC and whenever a change of transport channel mapping is necessary only the transport channel IDs are used to indicate which channel will be used. In message example 2.16 the uplink transport channel ID for RACH = '0' and the downlink transport channel ID or FACH = '0' are assigned to this particular connection.

Note: It will be noticed that transport channel IDs used in RRC and NBAP signalling messages belonging to same connection/call may have different values (+/−1) due to different number value ranges defined by 3GPP (for more details refer to Kreher and Ruedebusch, 2005).

In some cases PS radio bearers are also deleted and the UE is sent to RRC CELL_PCH state due to the long period of user inactivity on the user plane. In such cases a message like the one seen in message example 2.17 can be monitored.

Message example 2.17 RRC physical channel reconfiguration (RRC state = CELL_PCH)

ID Name	Comment or Value	
TS 29.331 DCCH-DL (2002-09) (RRC_DCCH_DL) physicalChannelReconfiguration		
dL-DCCH-Message		
2 message		
2.1 physicalChannelReconfiguration		
2.1.1 r3		
2.1.1.1 physicalChannelReconfiguration-r3		
2.1.1.1.1 rrc-TransactionIdentifier	0	
2.1.1.1.2 **rrc-StateIndicator**	**cell-PCH**	
2.1.1.1.3 utran-DRX-CycleLengthCoeff	5	
2.1.1.1.4 frequencyInfo		
2.1.1.1.4.1 modeSpecificInfo		
2.1.1.1.4.1.1 fdd		
2.1.1.1.4.1.1.1 uarfcn-DL	10761	
2.1.1.1.5 modeSpecificInfo		
2.1.1.1.5.1 fdd		
2.1.1.1.6 dl-InformationPerRL-List		
2.1.1.1.6.1 dL-InformationPerRL		
2.1.1.1.6.1.1 modeSpecificInfo		
2.1.1.1.6.1.1.1 fdd		
2.1.1.1.6.1.1.1.1 primaryCPICH-Info		
2.1.1.1.6.1.1.1.1.1 **primaryScramblingCode**	**123**	

This message contains not only the RRC state, but also information about the cell which was used by the UE last. If the UE changes this cell due to mobility it needs to perform the RRC Cell Update procedure towards the SRNC (cause = 'cell reselection'). If the UE is moving faster it needs to perform many cell updates and if it moves too fast we would see too many cell updates or the network would not be able to complete them successfully. For this reason RRC state URA_PCH was defined: the UE in this state does not need to perform an update each time a cell is changed, only if the UMTS registration area (URA) is changed.

A UMTS registration area (URA) is a cluster of cells. The number of cells belonging to the same URA is not limited by 3GPP standards.

Knowing this background a low RRC Cell Update Success Rate may indicate that state transitions form CELL_PCH to URA_PCH are not always performed when necessary. Prerequisite: network elements (UE, RNC) support URA_PCH state, which is not guaranteed especially for equipment brought to market during the early days of UMTS network deployment.

The problem with analysis of the RRC cell update procedure is that there is no defined failure message for this procedure. In some cases the network will react to the Cell Update Request with release of the RRC connection. This happens if RRC state machines on both sides of the connection (UE and RNC) are completely off the track, for instance due to unrecoverable errors in RLC AM transport blocks. However, in most cases it is expected to simply see no answer to the Cell Update Request and the UE will try again and again until timer T307 (see 3GPP 25.331) expires and the UE releases all information of the existing RRC context.

Another problem is that some state transitions reported by RNC statistics to higher level network management systems seem to be impossible when looking at 3GPP definitions. An example is necessary to explain this.

From Figure 2.40 it becomes clear that state transitions from CELL_PCH to CELL_DCH are actually not possible, but RNC statistics report such state transitions. The reason is that 3GPP 25.331 only reflects state transitions on the UE side. If the UE is in CELL_PCH state it is impossible to send any RRC message, because in CELL_PCH the UE can only be paged. To perform a necessary cell update procedure the UE must enter CELL_FACH first before it is able to send the RRC Cell Update Request message on RACH.

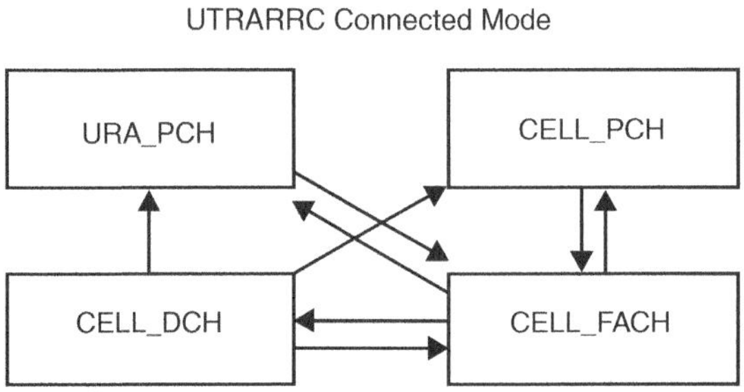

UTRARRC Connected Mode

Figure 2.40   RRC states as defined in 3GPP 25.331

The same thing happens if the UE is in CELL_PCH and wants to set up e.g. a voice call again. This means the UE needs to send an RRC Cell Update Request on RACH (cause = 'uplink data transmission'), but to be able to send this message it first needs to move from CELL_PCH state into CELL_FACH state. However, the SRNC is not informed about this UE internal state transition. From the point of view of the SRNC an RRC Cell Update message is received while the UE is still in CELL_PCH state and due to the fact that voice calls require CELL_DCH state the SRNC will order the UE subsequently to enter CELL_DCH state. The appropriate command is embedded in the RRC Cell Update Confirm message sent by the SRNC. This scenario explains why in RNC software and RNC statistics a state transition CELL_PCH to CELL_DCH exists although it does not exist in 3GPP specifications.

Performance measurement software based on Iub protocol analysis can only follow RRC messages and embedded RRC state indicators to detect state transitions. Due to performance reasons such software is also not able to store CELL_PCH state assigned to a number of UEs for a long time. The UE remains in CELL_PCH (or URA_PCH) state as long as the PDP context is active in the SGNS without transmitting either user data or signalling messages. A PDP context can be active for more than 48 hours without having an active RAB. If performance measurement software stores all connections with active PDP context, but no RAB (and also no message transfer between UE and RNC) active for more than 48 hours, it would need a lot of temporary memory resources that would be blocked for other temporary memory-intensive operations like macro-diversity filtering. For this reason most call trace applications will terminate the context of a call after the UE has been monitored being in CELL_PCH state for a couple of minutes. Also if performance measurement software is turned on at a defined time (maybe after hours or days of being turned off) it cannot retrieve any information from SRNCs about how many UEs in UTRAN are currently in CELL_PCH or URA_PCH state and hence are inactive from the point of view of transport resources. The only possible solution is to detect the start of new call/connection when RRC Cell Update Request (cause = 'uplink data transmission' or 'paging response') is transmitted from the UE to the SRNC.

Now – as discussed – the SRNC still shows RRC CELL_PCH or URA_PCH state for the particular UE when the RRC Cell Update Request message from this UE is received. By the appropriate Cell Update Confirm message the UE is already ordered to enter CELL_DCH state to be able to start voice calls and at this point of the call different state transitions are very clearly visible. While the UE will transit from CELL_FACH to CELL_DCH after receiving visible Cell Update Confirm message the SRNC has already registered a state transition form CELL_PCH (when visible RNC has received Cell Update from UE) to CELL_DCH (when RNC has constructed command that orders UE to enter CELL_DCH). To put it in a nutshell, if a procedure like RRC cell update is performed for a short time, RRC state machines on the RNC and UE side may not have the same state stored, because it is the nature of state machines that they need outgoing events (sent messages) and incoming events (received messages) to transit from one state to another.

2.11 CALL ESTABLISH SUCCESS AND FAILURE RATES

Basically it seems to be easy to calculate call establish success and failure rates, but there are some tricky options. It is important to measure the number of successfully established connections exactly, because it is an important input to call drop rates as well.

The first question that needs to be discussed is: What is the definition of a call in a UTRAN environment? According to chapter 1 a call is an active connection between the UE and the network used to exchange signalling and data. The exchange of signalling is a prerequisite to establishing a voice or data call. Hence, analysing the set up of signalling connections represented by SRBs is as important as analysing the establishment of voice and data connections represented by radio bearers (RBs) and/or radio access bearers (RABs).

2.11.1 RRC CONNECTION ESTABLISHMENT

Signalling radio bearers are established when an RRC connection between the UE and the SRNC is set up. This RRC signalling connection establishment is requested by the UE and executed/controlled by the SRNC as shown in Figure 2.41.

In this scenario an RRC connection setup success rate can be defined as follows. It does not matter if SRBs are mapped onto common transport channels (RACH/FACH) or dedicated channels (DCH).

$$\text{RRC Connection Setup Success Rate} = \frac{\sum RRC\ Connection\ Setup\ Complete}{\sum RRC\ Connection\ Request} \times 100\%$$

$$(2.21)$$

If the set up of the RRC connection fails there are three different cases. In case 1 the RNC is – due to load conditions – not able to offer a sufficient quality for the desired service in

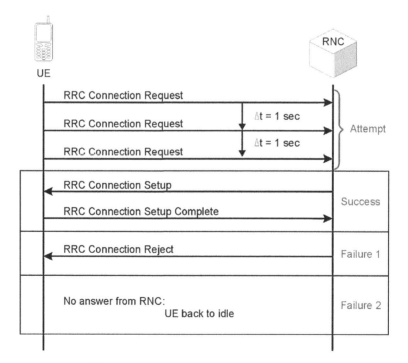

Figure 2.41   RRC connection setup procedure

the cell initially selected by the UE. Based on the establishment cause included in the RRC Connection Setup Request the RNC has already received the information about why signalling radio bearers are to be established: is it only to register to the network or does the UE wish to set up a voice or data call? If there are not enough resources available for the desired service the RRC connection establishment is blocked by the RNC that sends the RRC Connection Reject message. Rejecting the desired establishment of an RRC connection is also known as blocking.

The so-called RRC blocking rate is now defined as follows:

$$\text{RRC Blocking Rate} = \frac{\sum RRC\ Connection\ Setup\ Reject}{\sum RRC\ Connection\ Request} \times 100\% \qquad (2.22)$$

This RRC blocking rate is an important feedback for radio network planners, because typically they have designed network topology and availability of network resources in a way that the blocking rate per cell is approximately 1%. If in reality this value is higher or lower measurement results can help to optimise topology and resource planning.

If the set up of signalling radio bearers is blocked, the establishment of an RRC connection must not immediately fail. Depending on the feature availability in RNC software an RRC Connection Reject message may contain redirection information that is used to redirect the UE to a neighbour cell. This neighbour cell could work on a different UTRAN frequency or it is also possible to send the UE to GSM cells. The procedure is called RRC redirection. If there is no redirection information included in RRC Connection Reject the UE will perform cell reselection as described in 3GPP 25.304.

The second failure case is that the UE's RRC Connection Request is not answered at all although it is sent several times to the same RNC. The RRC connection setup procedure on the UE side is guarded by timer T300 and counter N300. The default value of T300 is one second, the default value of N300 is three. This means that if the RNC does not answer, the UE will send three RRC Connection Request messages using a time difference of one second (as shown in Figure 2.41). If all three attempts fail the UE falls back into IDLE mode.

In this case of multiple attempts the messages protocol analysis is confronted with the philosophical question of whether each unanswered RRC Connection Request is to be counted as a failed procedure or if only the final result of the procedure counts. If only the final result counts the overall success rate of the procedure is computed as 100%, if RRC Connection Setup and RRC Connection Setup Complete is monitored after the third RRC Connection Request sent subsequently by the same UE. Since there are different points of view even within operator service groups a final answer cannot be given in this book. Certainly there is no impact on the customer perceived quality of service if the UE needs to send more than one RRC Connection Request – it just takes one second longer to set up a call. On the other hand if this behaviour of network is monitored more often it must be guessed that something is wrong either in RNC software or in the transport network (because it is also possible that although RRC Connection Request has been monitored at a certain point it has not been received by the RNC, e.g. due to problems in ATM routers).

As another failure case it must be taken into account that the UE is not able to respond to the RRC Connection Request sent by the RNC, because it has again lost radio contact with cell. Since this failure happens relatively often a number of NEMs have implemented a

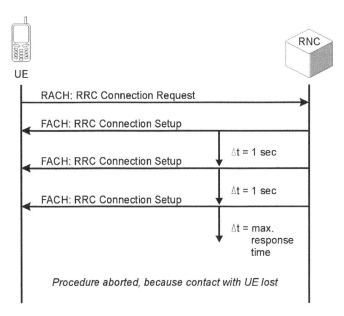

Figure 2.42 UE does not answer RRC Connection Setup

proprietary counter/timer function similar to the one in the UE in their RNC software. In a typical configuration scenario if the UE does not respond with RRC Connection Setup Complete within one second, then RRC Connection Setup will be sent again. If a second try is not successful either the RNC will send RRC Connection Setup a third time and then stop the procedure. From the perspective of performance measurement software the procedure must be detected as failed if after the second repetition of RRC Connection Setup plus waiting a typical maximum response time (not more than one second) there is still no answer from the UE as illustrated in Figure 2.42.

RRC Connection Setup 'no answer' failure rate can now be computed as follows:

$$\frac{\sum no\ Answer\ to\ RRC\ Connection\ Setup}{\sum RRC\ Connection\ Setup} \times 100\% \qquad (2.23)$$

There are no vendor-specific restrictions for this KPI. Timer and counter values described for the 'no answer' failure case might be different in that time periods are shorter and the maximum counter value is less. These timers and counters are not described in 3GPP standards.

Usually there are 10 times more 'no answer' failures measured in the case of RRC Connection Request establishment cause 'interRAT cell reselection'. A typical network scenario shows approximately a 10% RRC Connection 'no answer' rate for this particular establishment cause, while the same failure rate for all other establishment causes is typically measured in a range from 0.5 to 1% depending on the RNC manufacturer. Root causes are different Ec/N0 thresholds defined for the inter-RAT cell reselection procedure.

Knowing all failure cases the total RRC connection setup failure rate can be computed as follows:

$$\frac{\sum RRC\ Blocked + \sum RRC\ Request\ not\ answered + \sum no\ Answer\ to\ RRC\ Connection\ Setup}{\sum RRC\ Connection\ Request^*} \times 100\%$$

(2.24)

The RRC cell update procedure is performed in a similar way and also guided by the timer/counter as described for the RRC connection setup procedure. For this reason cell update success/failure rates are not explicitly described in this book. However, one important difference still needs to be mentioned: instead of an RRC Connection Complete message, an RRC UTRAN Mobility Information Confirm, RRC Physical Channel Reconfiguration Complete, RRC Transport Channel Reconfiguration Complete or RRC Radio Bearer Reconfiguration Complete message is used – depending on which kind of specific information was embedded in the previous RRC Cell Update Complete message that can be used e.g for the set up of dedicated transport channels/dedicated physical channels for transport channel type switching.

2.11.2 RADIO BEARER AND RADIO ACCESS BEARER ESTABLISHMENT AND RELEASE

Establishment of voice and data calls can be checked by analysing radio bearer setup or radio access bearer setup procedures. From the protocol point of view an RAB Establishment Request sent on Iu triggers the set up of the radio bearer on the Iub/Uu interface and a successful RB set up triggers the successful completion of RAB Establishment. For reasons of root cause analysis it makes sense to analyse both procedures (but do not add counter results). An analysis focused on the user perceived quality of service needs to look at only one procedure on one interface.

Regarding aggregation levels it makes sense to show the analysis of this procedure related to involved network elements: the UE, SRNC, MSC and/or SGSN. Aggregation on cell level must take into account that when RRC Radio Bearer Setup is sent the UE might already be in soft handover situation (in contrast to GSM where handovers are only necessary if traffic channels are already set up). This means that one and only one RRC radio bearer setup procedure is always performed, but identical transport blocks containing segments of the same involved RRC messages are often seen transmitted on multiple Iub/Uu interfaces simultaneously.

An Iub RRC Radio Bearer Setup message is monitored if a new RB is established between the UE and SRNC. In the case of successful establishment RRC Radio Bearer Setup Complete is monitored. If the UE cannot accept the assigned radio resources on the physical, transport channel or radio bearer level it will send a RRC Radio Bearer Setup Failure message (see Figure 2.43).

A single RRC Radio Bearer Setup message may contain establishment information for more than just one radio bearer. There are as many RRC radio bearer setup procedures seen on Iub as connections between the UE and network are set up that require a different quality of service. If for instance an active voice call becomes a multi-RAB call by the

---

*Customer option: each single message or sequences of up to N300 RRC Connection Request messages are counted.

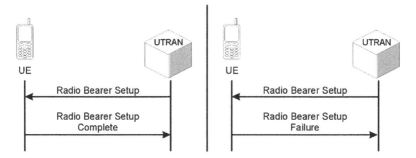

Figure 2.43   Successful and failed RRC radio bearer set up

establishment of another radio bearer to transmit IP payload in addition to voice packets it will result in an additional radio bearer setup for the same UE. Performance measurement software must ensure that such multiple procedures are shown correctly related to the same RRC connection. Success and failure rate formulas are quite easy to compute:

$$\text{RRC Radio Bearer Setup Success Rate} = \frac{\sum RRC\ RadioBearerSetup\ Complete}{\sum RRC\ RadioBearerSetup} \times 100\% \quad (2.25)$$

$$\text{RRC Radio Bearer Setup Failure Rate} = \frac{\sum RRC\ RadioBearerSetup\ Failure}{\sum RRC\ RadioBearerSetup} \times 100\% \quad (2.26)$$

Note: RRC reconfiguration procedures for physical channel, transport channel and radio bearer reconfiguration have identical request and failure messages and can generally be analysed in the same way as described above for RRC radio bearer setup.

Call flow and ratio formula definitions of the RRC radio bearer release procedure are similar to those described in case of the RRC radio bearer setup procedure if 'setup' is substituted by 'release' in the text. As its name says the RRC Radio Bearer Release Failure message indicates the failure of a radio bearer release on the UE side. This might not be a problem if subsequently RRC connection is terminated, but in the case of multi-RAB calls only one out of the multiple RBs will be deleted while others remain active. If RRC Radio Bearer Release Failure is received by the SRNC the new radio bearer configuration is deleted and the old configuration parameters are restored. It can be expected that a new RRC Radio Bearer Release message is then sent to the UE by the SRNC, which is another try to release the unused radio bearer. Separate counters for separate cause values monitored in the RRC Radio Bearer Release Failure message can be defined.

The analysis of RAB establishment (illustrated in Figure 2.44) is a little bit more difficult. The problem is that the same RANAP messages indicate both successful as well as unsuccessful RAB establishment.

The RANAP procedure is called RAB assignment and can be used 'to establish new RABs and/or to enable modifications and/or releases of already established RABs for a given UE' (3GPP 25.413). This means that if the RAB Assignment Request message is monitored this does not need to necessarily indicate the set up of a new RAB, it could also start the deletion of an existing RAB. Hence, a detailed analysis of information element sequences included in this message is necessary and attempt, success and failure protocol

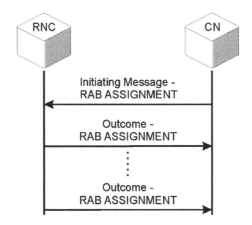

Figure 2.44 RANAP RAB establishment procedure

events cannot be counted based on the occurrence of RANAP messages. Message example 2.18 shows parameter details used to define the RAB establishment attempt counter:

The sequence that indicates an attempt event is id-RAB-SetupOrModifyItem and the number of counted attempt events must equal the number of included RAB-IDs, because there can be more than one RAB established. Actually it is also necessary to check if the RAB with this specific RAB-ID is already established and hence only modified, but in currently used network configurations RAB modification on RANAP level has not been observed so far. So this might become interesting as a future enhancement of performance measurement software.

Message example 2.18 RANAP RAB establishment attempt

ID Name	Comment or Value	
TS 25.413 V5.9.0 (RANAP) initiatingMessage (= initiatingMessage)		
ranapPDU		
1 initiatingMessage		
1.1 procedureCode	id-RAB-Assignment	
1.2 criticality	reject	
1.3 value		
1.3.1 protocolIEs		
1.3.1.1 sequence		
1.3.1.1.1 **id**	**id-RAB-SetupOrModifyList**	
1.3.1.1.2 criticality	ignore	
1.3.1.1.3 value		
1.3.1.1.3.1 sequenceOf		
1.3.1.1.3.1.1 sequence		
1.3.1.1.3.1.1.1 id	id-RAB-SetupOrModifyItem	
1.3.1.1.3.1.1.2 firstCriticality	reject	
1.3.1.1.3.1.1.3 firstValue		
1.3.1.1.3.1.1.3.1 **rAB-ID**	**'05'H**	
1.3.1.1.3.1.1.3.2 rAB-Parameters		

Message example 2.19 RANAP RAB establishment success

ID Name	Comment or Value	
TS 25.413 V5.9.0 (RANAP) outcome (= outcome)		
ranapPDU		
1 **outcome**		
1.1 procedureCode	id-RAB-Assignment	
1.2 criticality	reject	
1.3 value		
1.3.1 protocolIEs		
1.3.1.1 sequence		
1.3.1.1.**1 id**	**id-RAB-SetupOrModifiedList**	
1.3.1.1.2 criticality	ignore	
1.3.1.1.3 value		
1.3.1.1.3.1 sequenceOf		
1.3.1.1.3.1.1 sequence		
1.3.1.1.3.1.1.1 id	id-RAB-SetupOrModifiedItem	
1.3.1.1.3.1.1.2 criticality	ignore	
1.3.1.1.3.1.1.3 value		
1.3.1.1.3.1.1.3.1 **rAB-ID**	**'05'H**	

Note: it is a question of definition if the occurrence of the above described RANAP signalling sequence is also counted as an RAB establishment attempt if it is part of an RANAP Relocation Request message. Looking only at a specific Iu interface (the new SRNC after successful relocation/handover and appropriate MSC and/or SGSN) it is really a new RAB establishment procedure, but looking at the UE nothing is changed, because it is still the same RAB, only UTRAN interface involved in data transmission has been changed.

The appropriate success protocol event of RAB establishment is an RANAP Outcome message that often looks as shown in message example 2.19.

It is an RANAP Outcome message that contains the same RAB SetupOrModifiedList and the same RAB-ID as attempt event. If the successful establishment of RABs due to incoming relocation on a specific Iu interface is to be checked, the same information elements embedded in an RANAP Relocation Request Acknowledge message need to be counted as well.

The tricky thing is that the same RANAP Outcome message can also be used to report RAB establishment failure as shown in message example 2.20 for the same RAB-ID.

The same RAB-FailedList can also be embedded in an RANAP Relocation Request Acknowledge message. Another important point to mention is that if the RAB establishment attempt contained multiple RAB-IDs some of these RABs might be established successfully while others might fail, which means that a single RANAP message may contain both lists to indicate successful and unsuccessful RAB establishment in one step.

Formulas used to compute RANAP RAB establishment success and failure rates look similar to the ones used for RRC radio bearer setup. As we have seen the difficulty is not in the formula, but in ensuring that the right protocol events are counted.

The same statement is true for RAB release procedures that can be executed using the RANAP Iu-Release procedure, but the release of single RABs in a multi-RAB call can also be performed using the same RAB assignment procedure that was used to establish the RAB before as proven by message example 2.21.

Message example 2.20 RANAP RAB establishment failure

ID Name	Comment or Value	
TS 25.413 V3.6.0 (2001-06) (RANAP) outcome (= outcome)		
ranapPDU		
1 **outcome**		
1.1 procedureCode	id-RAB-Assignment	
1.2 criticality	reject	
1.3 value		
1.3.1 protocolIEs		
1.3.1.1 sequence		
1.3.1.1.1 **id**	**id-RAB-FailedList**	
1.3.1.1.2 criticality	ignore	
1.3.1.1.3 value		
1.3.1.1.3.1 sequenceOf		
1.3.1.1.3.1.1 sequence		
1.3.1.1.3.1.1.1 id	id-RAB-FailedItem	
1.3.1.1.3.1.1.2 criticality	ignore	
1.3.1.1.3.1.1.3 value		
1.3.1.1.3.1.1.3.1 **rAB-ID**	**'05'H**	
1.3.1.1.3.1.1.3.2 cause		
1.3.1.1.3.1.1.3.2.1 **radioNetwork**	**failure-in-the-radio-interface-procedure**	

Message example 2.21 RANAP RAB release attempt

ID Name	Comment or Value	
TS 25.413 V5.9.0 (RANAP) initiatingMessage (= initiatingMessage)		
ranapPDU		
1 initiatingMessage		
1.1 **procedureCode**	**id-RAB-Assignment**	
1.2 criticality	reject	
1.3 value		
1.3.1 protocolIEs		
1.3.1.1 sequence		
1.3.1.1.1 **id**	**id-RAB-ReleaseList**	
1.3.1.1.2 criticality	ignore	
1.3.1.1.3 value		
1.3.1.1.3.1 sequenceOf		
1.3.1.1.3.1.1 sequence		
1.3.1.1.3.1.1.1 id	id-RAB-ReleaseItem	
1.3.1.1.3.1.1.2 criticality	ignore	
1.3.1.1.3.1.1.3 value		
1.3.1.1.3.1.1.3.1 **rAB-ID**	**'05'H**	
1.3.1.1.3.1.1.3.2 cause		
1.3.1.1.3.1.1.3.2.1 **nAS**	**normal-release**	

The protocol sequence that indicates RAB Release is RAB-ReleaseList containing one or multiple RAB-IDs. Separate counters for different cause values found in RAB Release Attempt can be defined. As in the case of RAB establishment, the RAB release success event contains the same protocol sequence and RAB-ID as attempt. RAB release failure is indicated by the occurrence of RAB-ReleaseFailedList.

2.12 CALL DROP RATES

The definition of call drop rates also seems to be a philosophical discussion due to different definitions of 'call'. If a call is defined (as used in this book) as an active connection used to exchange user data (e.g. voice or IP packets) between the UE and the network, the existence of a call is bound to the existence of an RAB that consists of an RB and an Iu bearer. If either radio bearer or Iu bearer is dropped, RAB drops subsequently.

Radio bearers use radio links (physical radio channels). If a radio bearer is mapped onto a single radio link and this radio link goes out of order, e.g. due to synchronisation problems, the radio bearer and the call will be seen as dropped. However, there is a chance to reinstall radio links either if the UE and cell resynchronise themselves on the radio interface or if the UE performs a successful RRC re-establishment procedure as already explained in chapter 1. Due to this special nature of radio interface failures and channel re-establishment a special analysis is recommended as explained in more detailed in section 2.13. Thus a dropped call is defined as follows:

A dropped call is a dropped RAB.

This definition leads to the definition that a multi-RAB connection actually consists of two or more single calls. If the radio connection to the UE is dropped all RABs are dropped and the number of dropped calls for a particular UE rises by two or more. All in all such a definition reflects the user perceived experience rather than problems seen in the network.

At first sight it looks as if the biggest advantage of this definition is that call drop events can be detected relatively easily. An analysis of RAB release is enough. If release cause indicates an abnormal reason the call has been dropped. The drop can be originated either by UTRAN (including the SRNC) or by the core network (including foreign networks that might be involved in calls). If UTRAN has caused the problem the drop will be started by sending an RANAP Iu-Release Request or RAB-Release Request message, which triggers a subsequent RANAP Iu-release procedure or RAB-release. RAB release is executed using the same RAB assignment procedure used to establish RABs (compare message examples in the previous section). The difference between both procedures is that in the case of Iu-release the RANAP connection will be terminated and hence no RAB can remain active while in the case of RAB release only the RAB is terminated, not the Iu signalling connection between the SRNC and the core network element.

Note: it may happen that release causes in RANAP Iu-Release Request and the subsequent RANAP Iu-release procedure are different (as shown in Figure 2.45). Also if Iu-Release Request clearly indicates a failure on radio interface Iu-release may include cause value 'normal release', because this cause value refers to the release of RANAP dialogue only, not to the release of RABs.

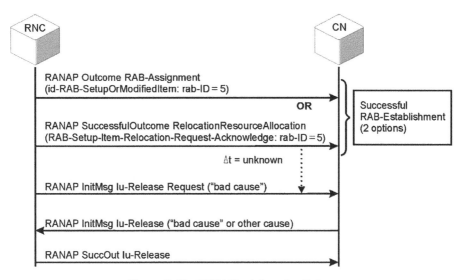

Figure 2.45 UTRAN originated call drop

To get the total number of dropped RABs it is necessary to count the occurrence of the following signalling events:

- **RANAP InitiatingMessage Iu-Release Request ('bad cause')**: single or multiple RABs of the same UE in the same CN domain are dropped on request by the RNC (UTRAN).
- **RANAP InitiatingMessage RAB-Release Request ('bad cause')**: single RAB of one UE in one CN domain is dropped on request by the RNC (UTRAN).
- **RANAP InitiatingMessage Iu-Release ('bad cause')**: single or multiple RABs of the same UE in the same CN domain are dropped by the core network (MSC or SGSN).
- **RANAP InitiatingMessage RAB-Assignment (RAB-Release List, 'bad cause')**: single RAB of the same UE in the same CN domain is dropped by the core network (MSC or SGSN).

Note: the number of counts must equal the number of RABs (influenced RAB-IDs) terminated by one of the above listed events.

There are two facts that make the correct counting of dropped RAB tricky: multi-domain calls and multi-RAB calls within a single domain.

Actually both of these 'call types' belong to a group of calls identified by the generic term 'multi-RAB' (as introduced in section 1.1.3), but for combined RAB signalling analysis it is necessary to specify more detailed subgroups. In the context of these subgroups a *multi-domain call* means a single UE having at least two active RABs, one RAB in CS and one RAB in PS domain. A *single domain multi-RAB call* means that one UE has multiple CS or multiple PS RABs active simultaneously. In each scenario of such a multi-domain or single domain multi-RAB call it is possible that one or all RABs in a certain domain are dropped by the core network or on request of UTRAN. A combination of all possible RAB drop scenarios is shown in Table 2.23.

Table 2.23 Possible cases/combinations of call drop scenarios

		Single domain call Single drop	Multiple drops	Multi-domain call Single drop	Multiple drops
UTRAN Originated drop	Iu-ReleaseRequest ('bad cause')	Case 1	Case 5	Case 7	Case 11
	RAB-ReleaseRequest ('bad cause', RAB-ID)	Case 2	N/A	Case 8	N/A
Core network originated drop	Iu-Release ('bad cause')	Case 3	Case 6	Case 9	Case 12
	RAB-Assignment (RAB Release List: 'bad cause', RAB-ID)	Case 4	N/A	Case 10	N/A

Case 1 mentioned in Table 2.23 has already been presented in Figure 2.45. Case 5 shows a variant of Case 1 that becomes valid if more than one RAB has been established in the same CN domain as illustrated in Figure 2.46.

In this case it becomes very obvious how Iu-release can drop multiple RABs simultaneously without indicating IDs of dropped RABs.

Case 7, which shows how single RABs in a multi-domain call are dropped on request by UTRAN, is illustrated in the call flow scenario of Figure 2.47. Please note that Iu-Release Request/Iu-Release messages do not contain RAB-IDs and hence it is necessary to look at correlated RAB establishment procedures to identify how many and which RABs are influenced by a single Iu-release.

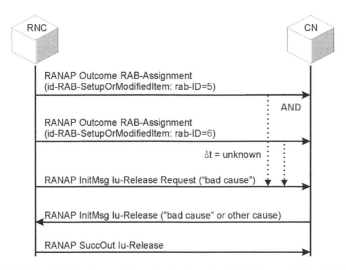

Figure 2.46 Multiple RABs dropped in single domain, RAB dropped by UTRAN, RANAP connection terminated

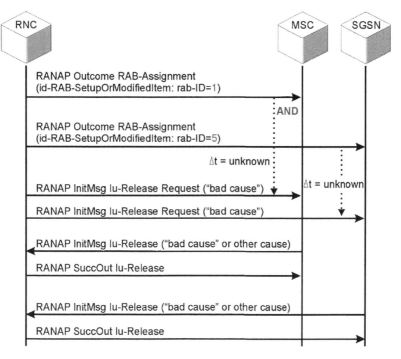

Figure 2.47   Single RABs dropped in a multi-domain call

This scenario will normally be seen if contact with the UE on the radio interface is lost while the UE has two active RABs, one in each CN domain. Cases 11 and 12 are variations of this case including the third RAB active in another CN domain.

The call drop scenario of a single drop in a multi-domain call in case 9 looks completely different (see Figure 2.48).

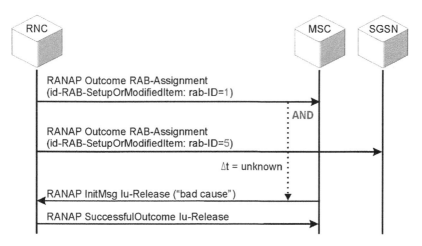

Figure 2.48   Multi-domain multi-RAB call, CS RAB dropped by core network

In this case the CS RAB is dropped while the PS RAB remains active. This scenario proves that the definition in which a dropped call must be counted as a dropped RAB is also valid if the user perceived experience is to be measured, because the lost speech call in the CS domain will be definitely recognised by the subscriber and despite continued PS connection he/she will claim that a connection with the network has been lost.

All remaining cases are variations/combinations of the above described.

RANAP 'bad cause' values that indicate dropped RAB are:

- unspecified-failure
- failure-in-the-radio-interface-procedure
- release-due-to-utran-generated-reason – this might not necessarily indicate a dropped call, because it is also used to indicate hard handover
- user-inactivity – a common value that indicates release of the RAB if the UE is sent to CELL_PCH, URA_PCH or IDLE mode while having an active PDP context. In these cases there is no dropped call, but if the same cause value is seen in the case of conversational or streaming calls it is a drop.
- iu-up-failure – indicates a transmission error occurred on the CS Iu bearer that can lead to the drop of the CS RAB
- radio-connection-with-UE-lost
- signalling-transport-resource-failure – indicates that the signalling connection is broken, which may lead to the subsequent drop of RAB(s).

There are many other RANAP cause values that indicate abnormal behaviour, but usually they indicate the unsuccessful set up of RABs or RANAP signalling, not the termination of already active connections. However, scenarios may vary depending on specific implementations of network equipment manufacturers.

So far a definition has been given of how to count the number of drops, however, to compute a drop rate, the number of active calls (= number of active RABs) on a defined Iu interface is also necessary. At first sight this seems to be equal to the number of established RABs (and this is what is usually assumed when call drop rates are computed in a laboratory environment), but looking deeper into the details of live network behaviour we recognise that the number of active connections is also determined by the number of incoming and outgoing handovers switching in the core network.

The number of active calls using a single Iu interface must hence be defined as follows (numbers in parenthesis refer to numbers used in Figure 2.49):

$$\sum \text{new established RABs (1)}$$

$$+ \sum \text{RABs established following incoming relocation/handover (2)}$$

$$- \sum \text{RABs released due to normal release of calls (3a)}$$

$$- \sum \text{RABs release due to outgoing relocation/handover (4)}$$

$$= \sum \textbf{Active calls (RABs)}$$

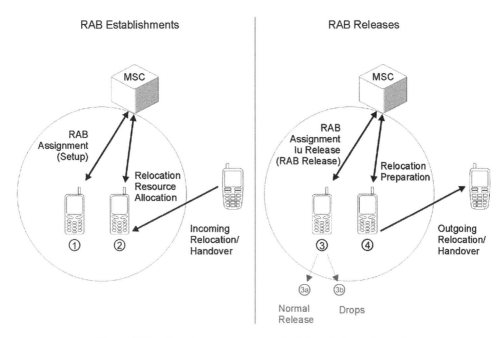

Figure 2.49 Counters necessary to calculate call drop rate on Iu

RABs are initially established using the RANAP RAB assignment procedure (1). RABs established due to incoming relocation/handover are established using the RANAP relocation resource allocation procedure. On the other hand RABs are released due to outgoing relocation/handover after an RANAP relocation preparation procedure (the release itself is usually executed by performing the Iu-release procedure). Remaining RABs in an area controlled by a single core network element (in Figure 2.49: MSC) are released using RAB assignment or Iu-release procedures (that might be triggered by Iu-Release Request). Based on the analysis of cause values the total number of released RABs (3) can be divided into cases of normal release (3a) and drops (3b).

If the number of active calls is known and the number of drops is counted exactly the call drop rate in % is defined as the number of dropped calls divided by the number of active calls multiplied by 100.

2.13 NBAP RADIO LINK FAILURE ANALYSIS AND RRC RE-ESTABLISHMENT SUCCESS RATE

As already discussed in the sections about hard handover analysis, the occurrence of an NBAP Radio Link Failure Indication message does not necessarily indicate a problem, because they are also often seen in the case of successful handover procedures. This is due to the fact that Node B detects more quickly that contact with the UE in a defined cell was lost than the handover target RNS or target RAT is able to signal successful relocation/ handover to source RNC that triggers more quickly radio link deletion procedure towards the source cell. In all other cases more quickly Radio Link Failure Indication message announces that one or more radio links or radio link sets are permanently unavailable and cannot be restored or that during the compressed mode uplink or downlink frames have occurred that had more than one transmission gap caused by a compressed mode sequence. These two cases are determined by the following cause values:

- synchronisation failure (permanently unavailable)
- invalid CM settings (multiple transmission gaps in single uplink or downlink frames)

Other cause values defined for this message indicate the following problems:

- Transport resources unavailable – indicates that a dedicated physical radio channel has become unavailable during an active connection.
- O&M intervention – contact with the UE lost due to interruption of network operator personnel.
- Control processing overload – indicates that Node B processor or software is in a critical situation, typically this is a software failure in Node B.
- HW failure – indicates hardware failure in Node B, e.g. due to power outage.

Most often the 'synchronisation failure' as shown in Figure 2.50 is monitored in live networks. Hence, it is useful to analyse this failure in a bit more detail. The root cause of this failure is that the synchronisation on the uplink dedicated physical channel of a radio link sent by the UE to Node B/cell has been lost. In other words, there is an interruption in uplink data transmission on a defined dedicated physical channel. And Node B has not been able to re-synchronise on this uplink path for a certain time period that is defined by timer $t_{RLFailure}$. This timer is specified in 3GPP 25.433 and the default value following the

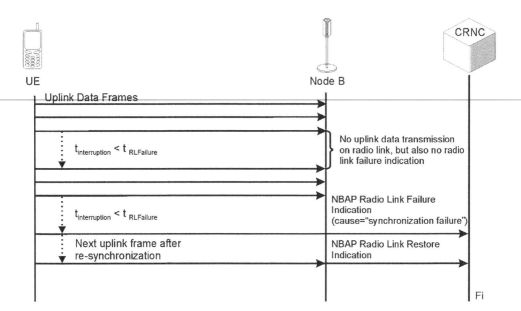

Figure 2.50 Interruption of uplink data transfer caused by synchronisation failure

NBAP specification is 10 seconds, but typically the configuration used by network operators is 5 seconds. Hence, if NBAP radio link failure indication with cause value 'synchronisation failure' is monitored this means that no uplink frame has been received on this radio link for 5 seconds. It also means that Node B has tried to re-synchronise with the UE for five seconds – without being successful. Now the CRNC is informed about this situation and it is up to the RNC to decide if an erroneous radio link will be deleted or not.

If there was just this single radio link in the active set of the UE and if there was no hard handover performed, the NBAP radio link failure indication could become the root cause of a dropped call. Indeed, the call must not necessarily be seen as dropped, because there are two possibilities of error recovery. If the radio link is not deleted it is possible (although unlikely) that the UE and the cell re-synchronise, which would be indicated by a new NBAP radio link restore indication belonging to the same call. Or it is possible that a UE that has lost radio contact with the network is able to re-establish its RRC connection as well as its dedicated physical channels after performing a cell update procedure – as already described in section 1.1.5.2.

If the UE was in soft or softer handover situation and only one radio link of the active set becomes unavailable it is possible that the RNC may decide to drop the call, but it is also possible that the call is continued using the remaining radio links. In the latter case it must be remembered what has been discussed in section 2.1.1 – that uplink BLER and hence the quality of higher layer services within a single cell is significantly influenced in soft hand-over situations, because the UE is located in the border area of a cell, but is not allowed to increase its transmission power. The higher number of block errors due to the increasing number of bit errors on the radio interface is either reduced by performing maximum ratio combining in Node B (softer handover) or quality selection combining, also known as macro-diversity combining in the RNC (soft handover). Now, if one radio link goes out of order the necessary input for combining functions (identical uplink frames transmitted on different radio links) is missed and this leads to a situation in which combining cannot eliminate the majority of bit and/or block errors anymore. Due to this the user perceived quality of service for this call will become significantly worse and the question will be asked: 'Wouldn't it be better to drop the call and let the UE have another try?'

Whenever radio link failures are monitored a root cause analysis is required. This can be done by correlating the time of occurrence of the radio link failure with either a radio quality parameter measured at the same time or protocol events known as possible error causes.

Radio contact with the UE, for instance, can be lost due to cell breathing effects. Such effects are caused by a rising interference in the uplink frequency band of the cell. The quality of the received signal falls below a critical threshold because of signals of UEs that are located in border areas of cell that are no longer able to increase their transmission power. Before the contact with the UE is finally lost the number of NBAP dedicated measurement reports showing extremely high SIR error and/or very bad SIR measurement increases. This could predict the upcoming situation. The overall noise level of the uplink frequency band is measured as the received total wideband power of the cell. Hence, if the occurrence of radio link failure(s) corresponds with extraordinary bad measurement results of the received total wideband power this could be the root cause of the lost radio contact.

Another possible root cause of synchronisation failures often observed in live networks is the successful activation of compressed mode. This problem is different from the 'multiple gap sequences in uplink frames' problem indicated by failure cause 'invalid CM settings'. If

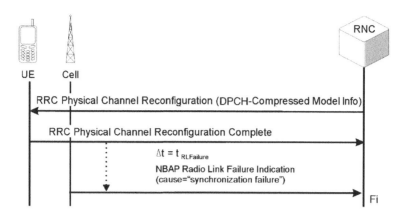

Figure 2.51   Radio link failure caused by compressed mode activation

compressed mode activation has caused a synchronisation failure, the connection with the UE was indeed lost due to compressed mode activation, which is very critical.

Note: use IMEI aggregation level/dimension to find out if such problems occur mostly when a certain type of UE is used. Use Node B aggregation level to find out if such problems might be caused by hardware or software of network equipment manufacturers.

To detect if the compressed mode activation has caused a radio link failure a very sophisticated analysis algorithm is required that cannot be based on simple counter values, because the value of $t_{RLFailure}$ must be part of analysis.

As shown in Figure 2.51 a request to the UE to activate compressed mode can only be identified by looking at the contents of the RRC Physical Channel Reconfiguration message. This message must include DPCH compressed mode information as highlighted in message example 2.22.

Message example 2.22 RRC physical channel reconfiguration request to activate compressed mode

I BITMASK IID Name	IComment or Value	I
I TS 25.331 DCCH-DL - V3.13.0 (RRC_DCCH_DL) physicalChannelReconfiguration (physicalChannelReconfiguration)		I
I dL-DCCH-Message		I
I		
I 2 message		I
I 2.1 physicalChannelReconfiguration		I
I 2.1.1.1.5 dl-CommonInformation		I
I 2.1.1.1.5.1 modeSpecificInfo		I
I 2.1.1.1.5.1.1 fdd		I
I 2.1.1.1.5.1.1.1 **dpch-CompressedModeInfo**		I
I 2.1.1.1.5.1.1.1.1 tgp-SequenceList		I
I 2.1.1.1.5.1.1.1.1.1 tGP-Sequence		I
I 2.1.1.1.5.1.1.1.1.1.1 tgpsi	I 4	I
I 2.1.1.1.5.1.1.1.1.1.2 tgps-Status		I
I 2.1.1.1.5.1.1.1.1.1.2.1 activate		I

If compressed mode activation fails (RRC Physical Channel Reconfiguration Failure is monitored) there will be no NBAP radio link failure indication due to compressed mode activation, because the counter is only increased if frame conditions of the failure indication as shown in Figure 2.51 are true. Once again this is a good example for a cumulative event counter not based on single protocol events, but on analysis of signalling protocol patterns.

For reasons of root cause analysis this counter can be displayed together with call drop rates/event counters derived from analysis of RANAP signalling. It also makes sense to correlate it with the analysis of RRC re-establishment procedure because in general UEs that have lost contact with network due to radio link failure may perform RRC re-establishment to get dropped radio links back into service.

The analysis of the RRC re-establishment procedure is a difficult thing because this procedure uses the RRC cell update procedure and if the cell update fails there is no distinctive protocol event. However, before looking at possible failure scenarios successful RRC re-establishment will be discussed.

As shown in Figure 2.52 the attempt event is an RRC Cell Update message that contains cause 'radio link failure'. Cell update messages are used for many purposes and only the 'radio link failure' cause indicates an attempted RRC re-establishment. The name of the procedure is actually not correct, it should rather be called radio link re-establishment or radio bearer re-establishment, because RRC entities continue their dialogue using common transport channels RACH and FACH after radio contact using dedicated physical channels has been lost. However, the radio bearer cannot switch easily to common transport channels and user plane data transfer is interrupted until dedicated radio links are in service again. To achieve this new radio link, a new Iub transport bearer to carry signalling radio bearers (SRB for RRC connection) and radio bearer (user plane data) need to be established as well. Then the UE receives a new dedicated transport channel configuration using an RRC Cell Update Confirm message that is sent via FACH. These parameters are similar to the ones usually

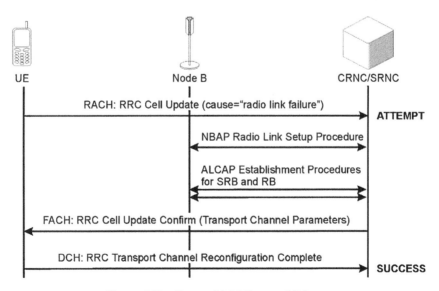

Figure 2.52   Successful RRC re-establishment

seen in RRC Connection Setup or RRC Radio Bearer Setup messages. Using an RRC Transport Channel Reconfiguration Complete message that has already been sent on the newly established radio link, the UE confirms that the re-establishement procedure has been successful.

On the other hand there are four failure cases that need to be taken into account.

1. Failure case 1: the UE sends multiple RRC Cell Update messages, but receives no answer from the RNC. If the maximum counter value for multiple attempt is reached the UE deletes all channel parameters still stored and goes back to IDLE mode. This is similar to RRC connection setup failure event 2.
2. Failure case 2: the RNC answers with RRC Cell Update Confirm, but there is no answer from the UE because radio contact with cell has been lost again. This is similar to RRC connection setup failure due to no answer from the UE although the RNC sent an RRC Connection Setup message up to three times.
3. Failure case 3: the UE receives RRC Cell Update Confirm, but due to an error in UE software the mobile cannot switch to the provided radio link. In this case an RRC Transport Channel Reconfiguration Failure will be monitored on RACH. This failure has no 'sister event' in the RRC connection setup scenario.
4. Failure case 4: in this case the re-establishment of radio links is rejected and more than this, the RNC decides to terminate the still active RRC connection with the UE by sending an RRC Connection Release message on common transport channel FACH.

It does not happen often that an attempted RRC re-establishment procedure is aborted, but it usually happens in the case of RRC cell update due to 'unrecoverable RLC error' indicated by the cause value with exactly this name as shown in Figure 2.53.

An unrecoverable RLC error only occurs if RLC acknowledge mode is used and despite the re-transmission of transport blocks that have been reported as erroneous the error cannot be recovered. Such behaviour is often observed in the case of UMTS UEs equipped with early generation chipsets while they are working in CELL_FACH state. For some reason they seem to be pretty sensitive to transmission errors or reduced link quality of S-CCPCH/ FACH. It is also seen as a rule that RRC Connection Release messages sent due to un-recoverable RLC errors mostly contain the cause value 'unspecified'.

Whether unrecoverable RLC errors or radio link failure might have been caused by bad radio link quality on the downlink frequency band can be investigated by analysing Ec/N0

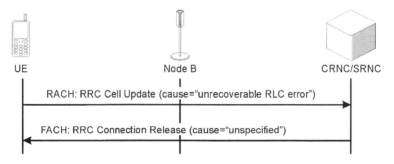

Figure 2.53 RRC connection terminated due to unrecoverable RLC error

values attached to RRC Cell Update messages. If these values do not fit into the normal distribution of Ec/N0 values extracted from RRC measurement reports a radio quality problem can be assumed, caused e.g. by an external interferer as shown in Figure 1.37.

2.14 CELL MATRICES

Cell matrices are a perfect tool to display an overview of radio-related measurement results found in RRC measurement report messages. Such measurement parameters in FDD mode are especially:

- chip energy over noise (Ec/N0);
- received signal code power (RSCP) – both measured on primary CPICH; and
- UTRA carrier RSSI.

Usually only results measured on the P-CPICH of intra-frequency neighbour cells are of interest. The matrix format is caused by a simple fact: if the UE sends an RRC measurement report containing radio quality parameter measurement results of multiple cells it means that the UE is located at a place where multiple cells overlap. Some of these cells might be members of the active set of the UE, others are just neighbour cells sending on the same frequency as the active set cells, but they are measured by the UE. Now it depends on the observer's point of view from which cell he/she looks at all other measured cells (see Figure 2.54). If, for example, a scenario with three overlapping neighbour cells, A, B and C, is defined and there is an observer who is looking at UE and its reported measurement results from cell A this observer will report: 'Cell B is measured at Ec/N0 level X while the UE is in cell A and cell C is reported at Ec/N0 level Y while the UE is in cell A. Finally cell A is also reported at Ec/N0 level Z while the UE is located in overlapping cell A.' This is a

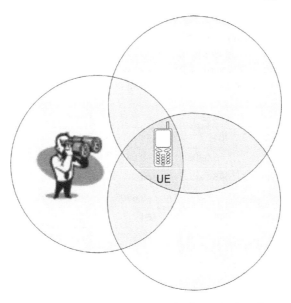

Figure 2.54 Watching measurement results from single cell perspective

Table 2.24   Abstract cell matrix with three different cells

	Cell A	Cell B	Cell C
Cell A	Ec/N0 = Z	Ec/N0 = X	Ec/N0 = Y
Cell B	Ec/N0 = Z	Ec/N0 = X	Ec/N0 = Y
Cell C	Ec/N0 = Z	Ec/N0 = X	Ec/N0 = Y

cell A orientated point of view. However, similar observers can also be imagined in cell B and cell C. They will report the same Ec/N0 levels for each cell, but the observer in cell B will say: 'Cell A is measured at Ec/N0 level Z while the UE is in cell B...' and so on.

Three different cells create three different points of view that can be imagined as three different lines in the cell matrix. Now three different measurement results that are also related to the three different cells must be reported for each line – the result of this presentation method is the cell matrix shown in Table 2.24.

Since in the case of a single RRC measurement report all cells of the matrix show the same measurement values for the same cell this might look strange, but the value of the cell matrix is only apparent if many measurement results have been monitored for each displayed cell. Then the different average values in matrix fields will show how much neighbour cells interfere with each other.

Note: due to the nature of WCDMA FDD mode cells only interfere with each other on the downlink frequency band. Uplink data transmission is not influenced by the interference visualised in cell matrices.

The example matrix in Figure 2.55 shows average Ec/N0 values extracted from data found in RRC measurement report messages. Each cell is identified by its primary scrambling code. The statement that can be derived if one looks at the marked matrix field is: 'Cell 336 was reported at an average signal level of -12.65 dB in all cases when the same RRC measurement report has also contained an Ec/N0 value for cell 448'.

Due to the fact that there are only 512 primary scrambling codes available for the whole UMTS radio network it is common that the same primary scrambling code is used in multiple cells controlled by the same RNC. To ensure that measurement results coming from different cells using the same primary scrambling code are not mixed the topology module

Scrambl. Code	464	448	336	304
304	−8,6905	−9,6471	−12,76	−19,78
Scrambl. Code	448	464	336	304
336	−7,8	−8,6905	−12,4079	−19,78
Scrambl. Code	448	464	336	304
448	−7,8	−9,8077	−12,65	−20,1176
Scrambl. Code	464	448	336	304
464	−8,6905	−10,9615	−13	−20,0714

Figure 2.55   Ec/N0 cell matrix example

Table 2.25 Calculation table for Little'i'-Matrix

Cells reported in RRC meas. reports	Primary scrambling codes	Number of RRC measurement reports containing other PSC in addition to PSC in table line				Number of RRC measurement reports containing PSC in table row
		448	304	464	336	
	448	—	17	13	30	30
	304	17	—	21	25	25
	464	13	21	—	21	21
	336	34	25	21	—	38

needs to detect correlation between the NBAP cell identifier (which is unique within UTRAN) and the primary scrambling code. 3GPP has also recognised this problem and defined an option to broadcast and report the cell ID together with the primary scrambling code in Release 5 standards of the RRC specification. However, UE and RNC manufacturers still do not seem to support this option.

A special case of cell matrix is the so-called Little'i'-Matrix, sometimes also called 'cell overlapping matrix'. This matrix delivers a % value of how much neighbour cells overlap. If implemented in performance measurement software based on UTRAN signalling analysis this Little'i'-Matrix is based on counters that count the occurrence of neighbour cells in RRC measurement reports without paying attention to measurement results of Ec/N0 or RSCP. The idea is simple: the more often two neighbour cells are reported together compared to the total number of all measurement reports including at least one of those two cells the more they overlap. Table 2.25 demonstrates this calculation.

In the matrix fields we see e.g. in the first line how often a measurement report for cell 304 is delivered together with the measurement report for cell 448 (17 times). In total 30 measurement reports for cell 448 have been received. Now these counter values are used to calculate the overlapping factor Little'i' for cell 448/304 overlapping. The calculation results are shown in Table 2.26.

Based on monitored measurement results it is estimated that 56.67% of the area of cell 448 is overlapped by cell 304 and it is estimated that 68% of the area of cell 304 is overlapped by cell 448 (line two, field one).

Table 2.26 Little'i' percentage value matrix

Cells reported in RRC meas. report	Primary scrambling codes	Overlapping factor of cells			
		448	304	464	336
	448	—	56,67%	43,33%	100,00%
	304	68,00%	—	84,00%	100,00%
	464	61,90%	100,00%	—	100,00%
	336	89,47%	65,79%	55,26%	—

Table 2.27   Handover matrix example

	Softer handover radio link addition success rate		Soft handover radio link addition success rate		Inter-frequency handover success rate	
Cell-ID	1002	1003	1011	1012	2001	2002
1001	100%	100%	100%	99.8%	99.4%	94.6%
Cell-ID	1001	1003	1011	1012	2001	2002
1002	100%	99.9%	100%	96.5%	98.2%	95.8%

Certainly this Little'i' matrix only delivers meaningful percentage values if there are a lot of measurement reports monitored for all cells. This can be achieved in three different ways:

- Monitor network over a long time period.
- Ensure that there are many UEs in monitored cells.
- Enable function to send periodical intra-frequency measurement reports to the SRNC.

A combination of those three ways is also possible, but one thing is for sure: all these matrices cannot substitute drive tests, they only support them.

Finally, there is another kind of very useful cell matrix that could be named 'handover matrix'. The idea is simple: if a defined number of UEs are located in cell A and they are moving into handover situations you want to find out to which neighbour cells they moved after successful handover or successful radio link addition in the case of soft handover. Instead of checking only success or failed handover cases such a matrix will deliver information about the distribution of source and target cells of the handover and hence it will allow statements about the preferred mobility directions of UEs. This kind of handover matrix can be used for all kinds of handover: intra-frequency, inter-frequency and inter-RAT handover.

As Table 2.27 shows the structure of the handover matrix must reflect the individual neighbour cell relations as well as different types of handover. It is possible to combine the analysis for different types of handover in one matrix as presented in the example or to display for each type of handover a separate matrix.

2.15   MISCELLANEOUS PROTOCOL PROCEDURES AND EVENTS THAT INDICATE ABNORMAL BEHAVIOUR OF PROTOCOL ENTITIES ON DIFFERENT LAYERS

This section contains a short description of some protocol events that indicate errors in different network elements on different layers. Usually these protocol events are not for in-depth analysis, but they are worth being recognised by performance measurement software especially for troubleshooting network elements including UEs.

The following subsections contain definitions of failure events. If not otherwise stated bold paragraph headlines are message names.

2.15.1 MISCELLANEOUS RRC FAILURE INDICATIONS AND RATIO KPIS

2.15.1.1 RRC UTRAN Mobility Information Failure

This message indicates that the UE cannot react to the previously received RRC UTRAN mobility command sent by the SRNC to allocate a new RNTI or to convey other mobility-related information. A failure message will include a cause value. Separate counters can be defined for separate cause values in this message. A failure ratio can be computed if the number of UTRAN mobility information failures is set in relation to monitored UTRAN mobility information. A success ratio can be computed as a ratio of UTRAN mobility information confirm messages divided by the number of UTRAN mobility information messages.

Note: it must be kept in mind that UTRAN mobility information confirm can also be sent by the UE after successful hard handover. In such cases the procedure code UTRAN mobility information is not used for an attempt protocol event. Instead any RRC Reconfiguration Complete or Cell Update Complete message must be counted as an attempt.

RRC UTRAN mobility information success rate =

$$\frac{\sum RRC\ UTRAN\ Mobility\ Information\ Confirm}{\sum RRC\ UTRAN\ Mobility\ Information} \times 100\% \qquad (2.27)$$

RRC UTRAN mobility information failure rate =

$$\frac{\sum RRC\ UTRAN\ Mobility\ Information\ Failure}{\sum RRC\ UTRAN\ Mobility\ Information} \times 100\% \qquad (2.28)$$

2.15.1.2 RRC Measurement Control Failure

This message indicates that the UE cannot initiate the RRC measurement procedure as requested by the SRNC. The failure message will include a cause value. Separate counters can be defined for separate cause values in this message. A failure ratio can be computed if the number of UTRAN measurement control failures is set in relation to the number of monitored RRC measurement control messages. It is not possible to compute a success ratio because there is no protocol message defined if that measurement is activated successfully. The only possible solution would be to look for the first RRC measurement report containing the same measurement ID information element as used in RRC measurement control, because if there are measurement results reported to the SRNC the initialisation of measurement procedure must have been successful.

$$RRC\ Measurement\ Initialization\ Failure\ Rate = \frac{\sum RRC\ Measurement\ Control\ Failure}{\sum RRC\ Measurement\ Control} \times 100\%$$

$$(2.29)$$

2.15.1.3 RRC Status

This message indicates a protocol error on the signalling connection between the UE and the network. The result can be a dropped call. Separate counters for separate cause values in this message can be defined, but no ratio KPIs can be computed because this is a unidirectional message.

2.15.1.4 RRC Security Mode Failure

This message indicates that the UE was not able to activate ciphering and/or integrity protection as required by the network (SRNC) in the previously received RRC Security Mode Control message. Separate counters for separate cause values in this message can be defined. The successful activation of ciphering/integrity protection is indicated by the RRC Security Mode Complete message. Hence, these protocol events can be used to compute ratio KPIs as follows:

$$\text{RRC Security Mode Success Rate} = \frac{\sum RRC\ Security\ Mode\ Complete}{\sum RRC\ Security\ Mode\ Control} \times 100\% \quad (2.30)$$

$$\text{RRC Security Mode Failure Rate} = \frac{\sum RRC\ Security\ Mode\ Failure}{\sum RRC\ Security\ Mode\ Control} \times 100\% \quad (2.31)$$

2.15.1.5 RRC Transport Format Combination Control Failure

The transport format combination control procedure is used to restrict or enable generally allowed uplink transport format combinations within the transport format combination set of a single connection. In the case of the monitored RRC Transport Format Combination Control Failure message the UE has not been able to accept settings required by the network as sent in the RRC Transport Formation Combination message by the SRNC. Based on protocol events a failure rate can be computed, but no success rate, because there is no explicit success message defined in 3GPP 25.331 for this procedure.

RRC transport format combination control failure rate =

$$\frac{\sum RRC\ Transport\ Format\ Combination\ Control\ Failure}{\sum RRC\ Transport\ Format\ Combination\ Control} \times 100\% \quad (2.32)$$

Note: changes of allowed transport format combinations may have an impact on the maximum theoretical throughput of transport channels as used in the transport channel usage ratio formula.

2.15.1.6 RRC Paging Response

This is not an error indication message, but the question often asked is how successful paging on RRC level can be detected. RRC paging response is needed not only to calculate a paging success rate, but also to measure the time difference between paging message and paging response.

Depending on the RRC state of the UE there are two different paging responses possible. If the UE is in IDLE mode it will answer using an RRC Connection Request. In this message the establishment cause has the same value as the paging cause in the previously received RRC Paging Type 1 message.

If the UE is in CELL_PCH or URA_PCH mode it will send RRC Cell Update including cell update cause value 'paging response' to the SRNC to request the re-establishment of dedicated radio resources. The paging message as well as the paging response will also have the same UE identifier, most likely IMSI, TMSI or P-TMSI.

The problem with paging analysis is identifying single paging requests, because for a single request, multiple RRC Paging Type 1 messages are sent (one in each cell that is assumed to be a possible location of the paged UE). Hence, a simple counter that counts the occurrence of paging messages does not work. Another challenge is to show paging success rates as well as response time measurement results aggregated on cell level, because large number of cells are used to send paging messages, but the paging response will be monitored only in the one cell that is the current location of the UE.

2.15.2 SCCP FAILURE ANALYSIS

All SCCP failures are aggregated on following levels:

- RNC;
- core network element (MSC/SGSN);
- UE.

2.15.2.1 Connection Refused (CREF)

It is sometimes seen that the UE sends incorrect initial NAS signalling messages. In such cases core network elements usually react with SCCP Connection Refused as shown in Figure 2.56.

Since there is no dedicated cause value for such reasons and in the case of detach procedures it is often correct to get a CREF as response to SCCP CR it is recommended to highlight all those CREF that are responses to CR if CR does not carry any of the following NAS messages: IMSI Detach Indication or GPRS Detach.

Occurrence of such events can be rated using the following KPI definition:

SCCP connection refused ratio =

$$
\frac{\sum SCCP\ Connection\ Refused\ without\ embedded\ Detach}{\sum SCCP\ Connection\ Request} \times 100\% \qquad (2.33)
$$

Figure 2.56 SCCP CREF used by the core network to reject incorrect NAS signalling

2.15.2.2 Inactivity Check Failure

If the inactivity timer expires before an answer from peer entity is received the SCCP connection will be terminated by sending the SCCP RLSD (Released) message containing release cause value 'expiration of receive inactivity timer'. The occurrence of this error will have fatal results for the existing RANAP connections and subsequently also for active RABs, which are controlled by the RANAP.

It is expected that in such failure cases the RNC reacts by sending the RANAP Reset Resource message that contains a cause value such as 'signalling transport resource failure'. Further RANAP reset procedures may follow releasing active RABs that have been influenced by losing the SCCP connection.

2.15.3 RANAP FAILURE ANALYSIS

2.15.3.1 RANAP Reset Resource

If this message is received by one of the RANAP entities the following 3GPP 25.413 'resources and references (i.e. resources and Iu signalling connection identifiers) associated to the Iu signalling connection identifiers indicated in the received message' will be released locally. The message is acknowledged by the peer entity by sending an RANAP Reset Resource Acknowledge message. A distribution of counters per possible cause value is as useful for root cause analysis as drill-down capability to the influenced/changed network elements and single connections.

2.15.3.2 RANAP Reset

This message is used to release affected RABs and to erase all affected references for the specific core network element or RNC node that has sent the RESET message. A distribution of counters per possible cause value is as useful for root cause analysis as drill-down capability to the influenced/changed network elements and single connections.

2.15.3.3 RANAP Overload

Congestions on signalling connections can be reported by sending an RANAP Overload message. The sender can be a core network node or RNC. It is expected that the receiving entity starts activating signalling traffic reduction mechanisms to overcome the situation. How such mechanisms work and how far they have been implemented in network nodes is beyond the scope of this book.

2.15.4 NBAP FAILURE ANALYSIS

All NBAP Class 1 use the same signalling procedure that contain a:

- *NBAP_Class1_Attempt*: NBAP Initiating message
- *NBAP_Class1_Success:* NBAP Successful Outcome message
- *NBAP_Class1_Failure:* NBAP Unsuccessful Outcome message

Using these events, success and failure ratio KPI formulas can be defined as already described in section 1.1.1. For each NBAP procedure separate counters/formulas will be implemented. When aggregating NBAP counters/formulas on cell level, values of NBAP c-ID will be used to display analysis results. In other words:

Often NBAP procedures prepare a subsequently executed handover to a target cell. Ensure that all such NBAP events are shown at target cell aggregation level, but not related to any cell (including the best cell) that is part of the active set when handover is attempted!

As recommended in the sections about handover analysis it is also useful to define subsets of counters within the same NBAP procedure to determine if e.g. an NBAP radio link setup procedure is used to establish dedicated transport resources for signalling radio bearers or to prepare soft or hard handover. An advantage of such definitions is a faster root cause analysis.

Some class 1 elementary procedures indicate availability and outage of network resources. This is true for the following procedure codes:

- Reset
- Cell Deletion
- Common Transport Channel Deletion
- Block Resource
- Unblock Resource

Resource outage time can be measured as the time difference between the Block Resource event and the Unblock Resource event. Cell outage time can be measured as the time difference between Cell Deletion event and the Cell Setup protocol event. In a similar way the outage time of common transport channels can be computed.

If performance measurement software could link the occurrence of these events to influenced connections (think about calls dropped due to a cell deletion procedure) it would be a powerful feature for root cause analysis of network problems.

In addition, NBAP analysis needs counters for the following NBAP class 2 elementary procedures. Since these are unidirectional messages without response no ratio KPI formulas can be defined:

- **Dedicated Measurement Failure Indication** – indicates that a desired radio quality parameter of a single UE connection will not be reported.
- **Radio Link Failure Indication** – as extensively discussed in section 2.13 and handover analysis sections.
- **Common Measurement Failure Indication** – indicates that a desired radio quality parameter of a single cell will not be reported.
- **Resource Status Indication** – does not necessarily indicate an error, but is always sent if cells or physical channels go out of order or if a cell changes its capabilities, e.g. if HSDPA support in a cell becomes unavailable. The included cause value must be analysed to identify abnormal behaviour.
- **Error Indication** – used to report protocol errors (e.g. semantic errors) found in incoming messages of connectionless NBAP class 2 elementary procedures back to the protocol entity that has sent the NBAP class 2 message.

2.15.5 RLC ACKNOWLEDGE MODE RETRANSMISSION RATE

Although this is not a protocol error the retransmission rate measured for RLC AM serviced will be mentioned, because it is very important for root cause analysis in the case of throughput problems. If transport channel throughput is high, but user perceived throughput is low a high retransmission rate on the RLC could be the root cause.

Retransmitted RLC AM transport blocks can be detected using two different methods. 3GPP 32.403 recommends checking bitmasks of RLC Status messages and count blocks that are marked in these bitmasks as not received. Manufacturers of performance measurement software have a second and probably better approach. We may remember that the architecture of this software requires an RLC reassember application. And to reassemble correctly on RLC level it is mandatory to recognise all retransmissions and wait until the last required transport block is received. This is the perfect opportunity to count the number of retransmitted frames and provide this number using an internal interface to the performance measurement module.

Root cause of high retransmission rates might not necessarily be located on the radio interface. It has also already been observed in live networks that transmission problems in ATM transport networks of UTRAN caused the most retransmissions on RLC level.

3

Call Establishment and Handover Procedures of PS Calls using HSDPA

The following chapter explains special performance measurement requirements for PS calls that use HSDPA. Differences in performance measurement definitions and KPIs compared to Release 99 network architecture/calls will also be discussed.

3.1 HSDPA CELL SET UP

The set up of HSDPA-capable Release 5 cells is differs from the set up of Release 99 cells in that some additional signalling procedures can be monitored. The audit procedure performed between Node B and CRNC as well as the set up of cells (see Figure 3.1) and common transport channels RACH, FACH and PCH is identical to the procedures seen when a Release 99 Node B or cell is taken into service. However, after a successful audit procedure (that might have been caused by a Node B reset) the Node B indicates which of its local cells are HSDPA capable using the NBAP Resource Status Indication message. Unfortunately, this message cannot be monitored while the cell is in service and thus reliable topology status information about which and how many cells of a network are HSDPA capable is only available in the RNC operation and maintenance centre (OMC). Making this information available to performance measurement software requires again sophisticated topology detection algorithms or a topology import/export interface between this software and the RNC.

After a successful NBAP cell set up and common transport channel set up for PCH, RACH and FACH(s) for all HSDPA-capable cells, an NBAP shared physical channel reconfiguration procedure is executed, which contains information about how many high-speed physical downlink shared channel (HS-PDSCH) codes are available in each cell. This means the maximum number of DL spreading codes with the fixed spreading factor 16 that can be combined for radio transmission of HS-PDSCH. Based on this number of combined spreading codes the maximum theoretical DL data transmission rate on the high-speed physical shared channel can be calculated.

UMTS Performance Measurement: A Practical Guide to KPIs for the UTRAN Environment   Ralf Kreher
© 2006 Ralf Kreher

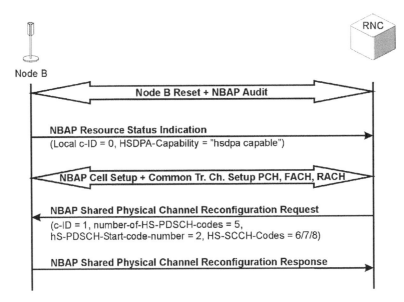

Figure 3.1   HSDPA cell set up

If there is more than one downlink spreading code per high-speed channel an HS-PDSCH start code number indicates which code will be used first. In addition to the HS-PDSCH information code the numbers of high-speed shared control channel (HS-SCCH) codes can be also found. By listening to these control channels, which are only sent on the radio interface, the UEs will receive information during the call and know which data packets, which have been sent on an HS-PDSCH, belong to which mobile.

After successful physical shared channel reconfiguration all necessary common resources to perform HSDPA calls are set up in a cell. A failure in this procedure would be reported using an NBAP Unsuccessful Outcome message since this procedure belongs to NBAP class 1 procedures.

3.2 HSDPA BASIC CALL

3.2.1 CALL SET UP AND MEASUREMENT INITIALISATIONS

A single UE using HSDPA works in the RRC CELL_DCH state. For downlink payload transport the HS-DSCH that is used is mapped onto the HS-PDSCH. The uplink IP payload is still transferred using the dedicated physical data channel (and the appropriate Iub transport bearer) known from Release 99 PS call scenarios. In addition, RRC signalling is exchanged between the UE and the RNC using the dedicated transport channels and the appropriate Iub transport bearer.

All these channels and bearers have to be set up and (re)configured during the call. In all cases both parties of radio connection, the cell and the UE, have to be informed about the required changes. While communication between Node B (cell) and the CRNC uses NBAP, the connection between the UE and the SRNC (physically the same RNC unit, but a different protocol entity) uses the RRC protocol.

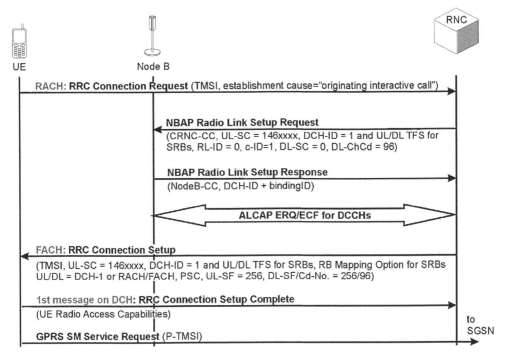

Figure 3.2   HSDPA call flow example 1/5

There is no rule that defines which traffic class a HSDPA call must have – except that it needs to be a traffic class used for PS calls. Therefore the call may start as any other web-browsing call using interactive or background traffic calls of UTRAN. The example calls shown in this section use the interactive traffic class that is more delay sensitive than the background/interactive.

First the UE sends an RRC Connection Request message to the RNC as illustrated in Figure 3.2. This triggers the NBAP radio link setup procedure after the admission control function of the RNC has decided to map SRBs onto the dedicated transport channels and orders the UE to enter CELL_DCH state. The NBAP Radio Link Setup Request message contains a CRNC communication context (CRNC CC), which is the identifier of the UE within the CRNC and it is interesting to track the initiation of NBAP dedicated measurements belonging to this call (but not shown in call flow). The message also contains the uplink scrambling code (UL-SC) assigned to the mobile: the unique UE identifier on the FDD radio interface. Furthermore, the DCH-ID and transport format set (TFS) settings of the DCH are included. They carry the signalling radio bearer, the radio link ID (rL-ID), the cell-ID (c-ID) of the NBAP cell, the downlink scrambling code and the downlink channelisation code belonging to the dedicated physical (downlink) channel. The NBAP radio link setup response (Successful Outcome message of this procedure) contains Node B communication context and the binding ID related to the previously signalling DCH-ID. This binding ID allows linking of the following ALCAP establishment procedure and the established Iub transport bearer (AAL2 SVC) to the radio link established for a particular UE.

RRC connection setup transmits the same physical channel and transport channel parameters to the UE as found in the NBAP radio link setup procedure. Now both the UE and Node B/cell are ready to communicate with each other using the dedicated physical radio channels. In uplink as well as in downlink these channels use spreading factor 256, which is typical for standalone signalling radio bearer transmission via DCH. The UE confirms that the dedicated channels have been successfully taken into service by sending an RRC Connection Setup Complete message to the SRNC.

Both the NBAP radio link setup and RRC connection setup procedure can be analysed using the same success and failure ratios as discussed in sections 2.15.1.4 and 2.11.1.1. There is no specific information to indicate a possible later usage of HS-DSCH at this stage.

The same is true for the following GRPS Session Management Service Request message sent by the UE to the SGSN to start PS call set up (see Figure 3.3). This message usually contains P-TMSI as a temporary UE identity.

After RRC connection has been established successfully an RRC Measurement Control message is sent by the SRNC to initialise intra-frequency measurement on the downlink frequency band in the UE. Usually this message contains a list of event triggers to be reported to the SRNC if the defined measurement options (such as hysteresis and time-to-trigger) are met. Events 1A, 1B and 1C are reported if the UE wishes addition, deletion or substitution of radio links to/from its active set. All these events belong to the group of soft and softer handover procedures. As it will be explained later in this chapter a soft or softer handover can also trigger the change of the serving HS-DSCH cell. For this reason this measurement initialisation is important although it is not different from the same message used in the Release 99 environment.

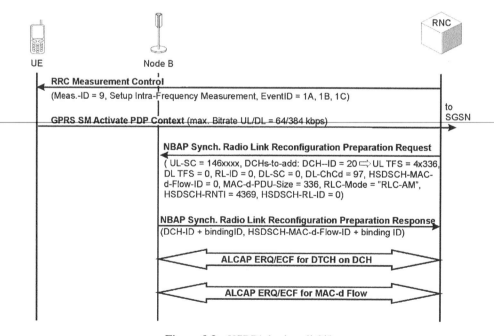

Figure 3.3   HSDPA basic call 2/5

While RRC measurements are initiated the UE continues the NAS signalling dialogue with the SGSN and requests the activation of a PDP context. The appropriate GPRS session management message also contains the desired maximum bit rates for uplink and downlink radio transmission. The 384 kbps downlink transmission rate shown in the call flow example does not necessarily need to trigger the assignment of HS-DSCH resources to this call. This decision is up to RNC's admission control function, because identical bit rates are found in the RANAP RAB Assignment Request message that triggers the subsequently monitored NBAP synchronised radio link reconfiguration preparation procedure on Iub. The request message of this NBAP procedure contains a list of DCH-IDs and the appropriate transport format set (TFS) parameters to be added. The DCH with DCH-ID = 20 shown in the example call trace (Figures 3.2 to 3.6) will be used to transmit the uplink IP payload while its downlink TFS is 0. This means that no downlink data transfer is possible on this DCH. Radio link ID (RL-ID), downlink scrambling code (DL-SC) and downlink channelisation code (DL-ChCd) are further parameters which are necessary to identify this dedicated radio link on Uu. The parameter definition to MAC-d flow follows. A MAC-d flow is a physical Iub transport bearer used to transport all downlink IP payload to be sent from the SRNC via HS-DSCH to Node B. Packet scheduling and multiplexing is done in Node B as well as hybrid ARQ error detection and correction. The throughput of a single or all MAC-d flows towards a defined cell can be measured in a similar way as described for the transport channel throughput measurement in Chapter 2, but transport channel throughput of the HS-DSCH cannot be measured by monitoring MAC-d flows, only on the radio interface directly. For this reason the reporting of the HS-DSCH provided bit rate using NBAP common measurement reports has been defined in Release 5.

If the HS-DSCH-provided bit rate shows a high value, but downlink throughput measured on all MAC-d flows to a defined cell seems to be too low it is recommended to look at the RLC AM retransmission rate. If the RLC AM retransmission rate also does not show critical values, the root cause of problems must be assumed to be located on the radio interface. Unfortunately the retransmission rate of HARQ is not reported by Node B.

An important parameter of the UE is HS-DSCH-RNTI (often called, H-RNTI). Using this ID, the UE is able to select its transport blocks from the HS-DSCH that is used by several UEs simultaneously. The HS-DSCH RL-ID (radio link ID) becomes important if more than one HS-DSCH is bundled in a HS-DSCH radio link set. A cell can have more than one HS-PDSCH and hence multiple HS-DSCHs per UE are possible.

In the NBAP Synchronized Radio Link Reconfiguration Response message some binding IDs can be seen that are used to tie appropriate Iub transport bearers (VPI/VCI/CID) to the DCH used for UL IP payload transmission and to MAC-d flow. These transport bearers are established using the well-known ALCAP procedures.

In the RRC Radio Bearer Setup message (see Figure 3.4) H-RNTI is found again in addition to the RRC state indicator that orders the UE to remain in CELL_DCH state (ASN.1 encoded: 'cell-DCH'). In the example call trace RAB, identified by RAB-ID=5, is represented by the radio bearer with RB-ID = 6 and this radio bearer has a number of mapping options.

Note: it is interesting that some network equipment manufacturers have defined that a call is in (virtual) 'HSDPA Active' state as long as the last mapping option DL transport channel = HS-DSCH is valid. This is true as long as the UE is located in HSDPA-capable cells, no matter if the HS-DSCH is used for downlink payload transmission or not. Only if the UE is handed over to a Release 99 cell (no HSDPA possible) this mapping option is

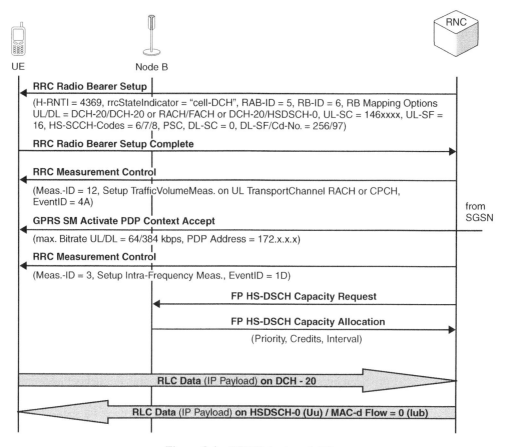

Figure 3.4   HSDPA basic call 3/5

deleted from the list of possible options using an RRC transport channel reconfiguration
procedure. On the other hand network operators want to see how often the HS-DSCH found
in RB mapping options is really used. In both cases state transitions may have the same
name, but completely different definitions.

The uplink scrambling code (UL-SC) will also remain the same for the UE if it is in
'HSDPA Active' state. The UL spreading factor (UL-SF) is 16, which is the standard spreading
factor defined for HSDPA. It actually allows uplink data transfer rates of 128 kbps although
only 64 kbps are required by the UE. In addition, code numbers are included of the provided
HS-SCCHs, the primary scrambling code (PSC) of cell and the downlink scrambling code
(DL-SC). Downlink spreading factor 256 allows only the signalling radio bearer (RRC
messages) to be transported on the dedicated downlink physical channel (DPCH), but the DL
channelisation code number is changed compared to the one used for RRC connection set up.

Having set up the radio bearer successfully another RRC Measurement Control message
initialises traffic volume measurement for the uplink RLC buffer in the UE that is relevant
for data to be transmitted on the RACH or common packet channel (CPCH). However, as
long as the CPCH is not found in radio bearer mapping options this channel does not need to
be considered to be involved in the call.

With the PDP Context Activation Accept message, the SGSN signals to the UE that the desired PS connection is now available. Independent from this NAS, the signalling RNC initialises another measurement procedure in the UE. This time intra-frequency event 1D is set up to be reported if conditions are fulfilled. 1D stands for 'change of the best cell' and will trigger HS-DSCH cell change as well (see the next section). After this, the last important measurement for mobility management of the call is defined. Meanwhile successful radio bearer establishment triggered some alignment procedure on the Iub frame protocol (FP) layer between Node B and RNC. Especially the capacity allocation procedure contains important performance-related parameters. Since HS-DSCH resources are controlled by Node B the RNC asks for provision of necessary transport resources using the FP HS-DSCH capacity request message. In the capacity allocation message Node B signals what was granted. RLC PDUs transmitted for this specific call are prioritised using a priority indicator that has a value in the range from 0 to 15 (15 indicates highest priority). The credit information element represents the number of RLC PDUs that the RNC is allowed to transmit within a certain time interval. The interval is a multiple of the TTI used in the HS-DSCH, a typical TTI value is 2 ms. When performance-related data from Node B is available together with the data captured by the Iub porbes it can be proved that there is a linear correlation between CQI sent by UE on radio interface to the serving HS-DSCH cell and the credits sent by the cell's Node B to the SRNC. Hence, evaluation of credits allows to estimate values of CQI although CQI is not transmitted via Iub interface. After all data has been transmitted by the RNC, Node B will send anther capacity allocation message that starts the next sequence of DL data transfer and this procedure will continue in the same way as long as HS-DSCH is used by a UE. Data transmission starts after each capacity allocation on Iub: uplink data is transported in the Iub physical transport bearer of the DCH and downlink IP data is transmitted in the AAL2 SVC of MAC-d flow, which will be multiplexed onto the HS-DSCH by Node B.

3.2.2 CALL RELEASE

A call release of an HSDPA call can be triggered by different events. As the usage of HS-DSCH is only one possible mapping option of a dedicated traffic channel (DTCH), which carries the IP payload, we can define that the HSDPA call is finished if the HS-DSCH is no longer used. A special case of HSDPA transport channel type switching is seen if during an active HSDPA call a voice call is attempted. In this case the HS-DSCH shall not be used any longer. Instead the downlink PS radio bearer of the connection is mapped onto a 384 kbps DCH as long as the voice RAB is active. Once the voice call is finished the downlink PS traffic channel will he mapped onto HS-DSCH again. The call flow example presented in the Figure 3.5 does not show such transport channel type switching. Instead it shows how the deactivation of the PDP context requested by the UE and executed by the SGSN terminates the radio connection and thus also the usage of the HS-DSCH.

After the SGSN has sent a Deactivate PDP Context Accept message it starts the RAB release procedure on IuPS (not shown in the figure), which subsequently triggers another NBAP synchronised radio link reconfiguration procedure. In this message we find a DCHs-Delete-List for the previously installed transport channels of the DTCH: a DCH for uplink IP payload transmission and an HS-DSCH MAC-d flow for downlink IP payload transport.

After these two transport channels have been removed from Node B/cell they need to be deactivated in the UE as well. For this reason the RRC Radio Bearer Release message is sent

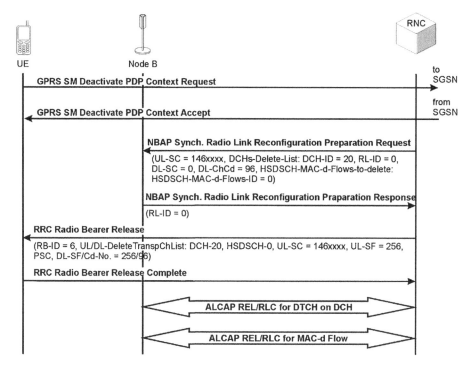

Figure 3.5   HSDPA basic call 4/5

to the UE. The appropriate RRC Radio Bearer Release Complete message confirms that all 'traffic channels' between the UE and the network have been deleted. Only SRBs for transmission of RRC and NAS signalling remain active for a while.

Since the DCH and MAC-d flow have been deleted, their AAL2 SVCs (VPI/VCI/CID) can be deleted as well. This is done using ALCAP release procedures.

Two subsequent RRC Measurement Control messages (Figure 3.6) terminate the PS call specific measurements and reporting events: 4A and 1D. The link to measurement initialisation is given by the identical measurement ID.

Now the RNC terminates the RRC connection, which means that the SRBs are deleted, too. This is confirmed by the UE as well and after finishing the connection on the radio interface the radio link in the cell and the appropriate Iub transport bearers for the dedicated control channels (DCCHs) are deleted. This is the last signalling event of this call.

3.3 MOBILITY MANAGEMENT AND HANDOVER PROCEDURES IN HSDPA

Mobility definitions for HSDPA are very strict. 3GPP has defined that HSDPA is only possible if the UE is in CELL_DCH state. There is one – and always only one – serving HS-DSCH cell. Usually this cell should be the best available cell of an active set, but this is not always guaranteed. Within the serving HS-DSCH cell one or more HS-PDSCH can exist. They all belong to the same serving HS-DSCH radio link of a UE. RRC connection mobility management is allowed to be realised by mobile evaluated soft and hard handover

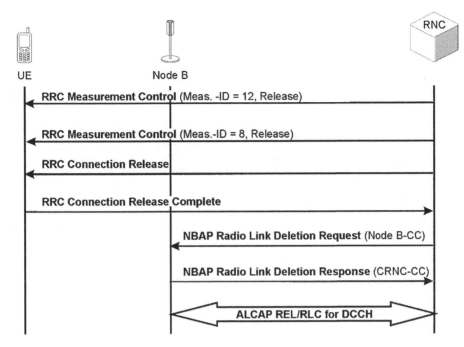

Figure 3.6 HSDPA basic call 5/5

procedures only. This means that serving HS-DSCH cell change – as HSPDA handovers are called by 3GPP – triggered by periodical measurement reports are not allowed.

3.3.1 SERVING HS-DSCH CELL CHANGE WITHOUT CHANGE OF ACTIVE SET

This handover scenario (see Figure 3.7) is also known as intra-node B synchronised serving HS-DSCH cell change. In this scenario the dedicated transport channels for uplink IP traffic and RRC signalling remain unchanged. The target-serving HS-DSCH cell is in the same

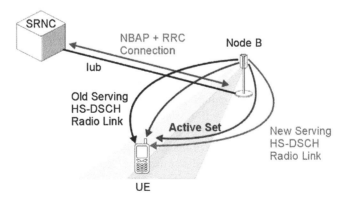

Figure 3.7 Intra-Node B synchronised serving HS-DSCH cell change overview

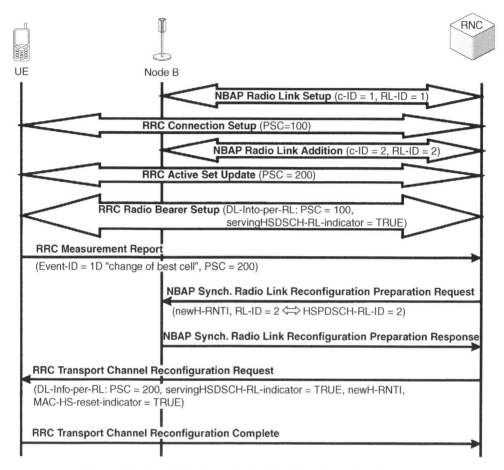

Figure 3.8 Intra-Node B synchronised serving HS-DSCH cell change

Node B as the source-serving HS-DSCH cell, which looks similar to a softer handover, but indeed there is no such handover because the active set of the UE remains unchanged.

The true softer handover procedure as shown in Figure 3.8 may have happened before radio bearer setup, which means before this call entered virtual 'HSDPA active' state. For an application that tracks the changing location of the UE it is important to analyse the initial NBAP radio link setup procedure as well as the following radio link addition procedure, because in these procedures cell identities (c-ID) have been signalled. From parameters found in the RRC radio bearer setup request message it emerges that the initially serving HS-DSCH cell is cell identified by PSC = 100, which is correlated to NBAP c-ID = 1.

The HSDPA handover is then triggered by the RRC measurement report containing event-ID 1D 'change of the best cell' and the primary scrambling code of the target cell.

When the SRNC has performed the handover decision, Node B is prepared for the serving HS-DSCH cell change at an activation time. This activation time is once again represented by a certain CFN. Another important parameter is the new H-RNTI that is assigned to the UE due to cell change.

Note: the new H-RNTI distinguishes RRC reconfiguration procedures used for HSDPA mobility management from RRC reconfiguration procedures with a different purpose (for instance inter-frequency hard handover).

An NBAP synchronised radio link reconfiguration preparation procedure is executed. Besides the new H-RNTI this message contains an HS-DSCH radio link id (HS-DSCH-RL-ID) that is correlated with the radio link id (RL-ID) already monitored in the NBAP radio link addition request message. On behalf of this correlation the new serving HS-DSCH cell is identified on the NBAP layer. After this procedure all necessary information regarding the reconfiguration is available in Node B. The SRNC then sends either a Physical Channel Reconfiguration message (as described in 3GPP standards) or a Transport Channel Reconfiguration messages (as introduced by some NEM and shown in figure 3.8), which indicates the target HS-DSCH cell identified by its primary scrambling code and the activation time to the UE. Since the same Node B controls both the source and the target HS-DSCH cells, the Iub transport bearer and its related parameters are not changed on the Iub interface. When the UE has completed the serving HS-DSCH cell change it transmits a Physical/Transport Channel Reconfiguration Complete message to the network.

If radio bearer parameters are changed, the serving HS-DSCH cell change needs to be executed by a radio bearer reconfiguration procedure, respectively.

3.3.2 INTER-NODE B SERVING HS-DSCH CELL CHANGE

The next mobility scenario (Figure 3.9) is a synchronised inter-Node B serving HS-DSCH cell change in combination with hard handover. The reconfiguration is performed in two steps within UTRAN. On the radio interface only a single RRC procedure is used.

It is expected that this procedure is performed especially if the new serving HS-DSCH cell is connected to a different RNC and no Iur interface is available or if the new cell is working on a different UTRAN frequency than the old one. Hence, there might be additional network elements and interfaces involved in this procedure that are not shown in this particular example of the scenario.

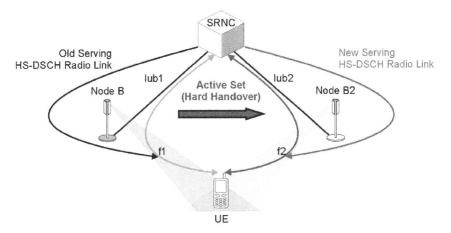

Figure 3.9   Inter-Node B synchronised serving HS-DSCH cell change overview

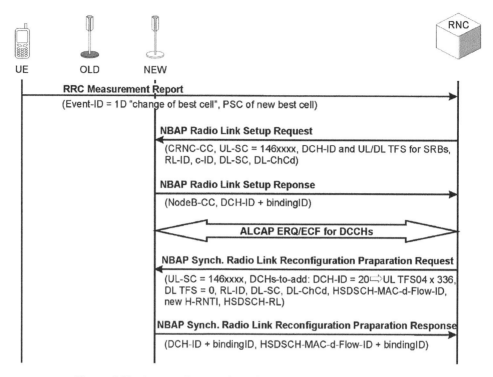

Figure 3.10   Inter-Node B serving HS-DSCH cell change call flow 1/2

The cell change is triggered once again when the UE transmits an RRC Measurement Reprot message containing intra-frequency measurement results, here assumed to be triggered by the event 1D 'change of the best cell' (see Figure 3.10). When the SRNC has performed the handover decision, Node B is prepared for the serving HS-DSCH cell change at an activation time, again represented by a certain CFN.

As an alternative to the RRC measurement reports the SRNC may also determine the need for a handover based on load control algorithms. Since the target cell is working on a different UTRA frequency than the source cell RRC measurements it might be necessary to activate the compressed mode before measurement is started.

In the first step, the SRNC establishes a new radio link in the target Node B. In the second step this newly created radio link is prepared for a synchronised reconfiguration.

The SRNC then sends a Transport Channel Reconfiguration message on the old configuration. This message indicates the configuration after handover, both for the DCH and HS-DSCH. The Transport Channel Reconfiguration message (Figure 3.11) includes a flag indicating that the MAC-d flow parameters and the appropriate Iub transport resources need to be established on the new configuration and to be released on the old Iub. The message also includes an update of transport channel related parameters for the HS-DSCH in the target HS-DSCH cell.

The UE terminates transmission and reception on the old radio link if the indicated CNF is received and configures its physical layer to begin reception on the new radio link. After radio link data transmission is synchronised, the UE sends a Transport Channel Reconfiguration

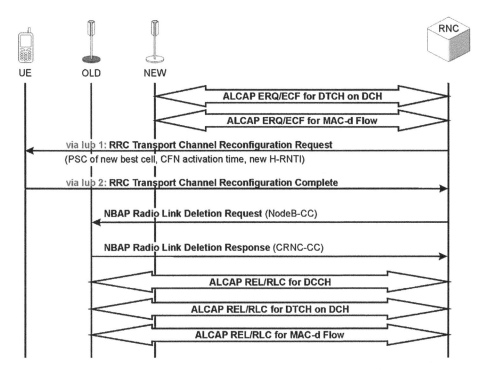

Figure 3.11   Inter-Node B serving HS-DSCH cell change call flow 2/2

Complete message. The SRNC then terminates reception and transmission on the old radio link for the dedicated channels and releases all resources allocated to the considered UE.

Note: in this inter-Node B handover example, the RLC for transmission/reception on the HS-DSCH is stopped at both the UTRAN and UE sides prior to reconfiguration and continued when the reconfiguration is completed.

3.3.3 HSDPA CELL CHANGE AFTER SOFT HANDOVER

The official name of this procedure in 3GPP standards is inter-Node B synchronised serving HS-DSCH cell change after active set update, but the scenario (Figure 3.12) is simply characterised by the fact that a change of the best cell following a successful soft handover subsequently triggers the executed serving HS-DSCH cell change. There is no separate measurement event sent to trigger the HSDPA mobility operation.

The active set update procedure may be used for either soft or softer handover.

In the illustrated example it is assumed that a new radio link is added, which belongs to a target Node B different from the source Node B. The cell, which is added to the active set, is assumed to become the serving HS-DSCH cell in the second step. This combined procedure comprises an ordinary active set update procedure in the first step and a synchronised serving HS-DSCH cell change in the second step.

In the example given in Figures 3.13, 3.14 and 3.15 it is assumed that the UE transmits an RRC Measurement Report message containing intra-frequency measurement results and event-ID 1A. Based on this measurement report the SRNC determines the need for the

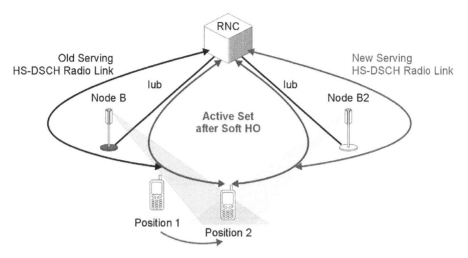

Figure 3.12   Inter-Node B synchronised serving HS-DSCH cell change after active set update overview

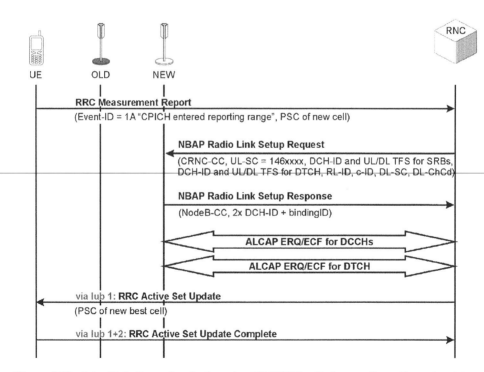

Figure 3.13   Inter-Node B synchronised serving HS-DSCH cell change after active set update call flow 1/3

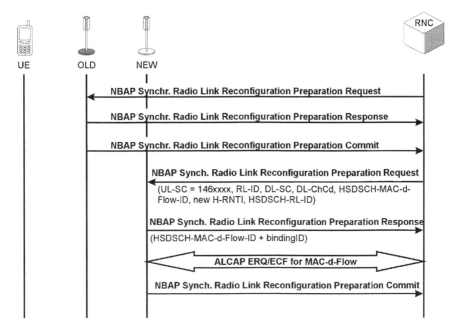

Figure 3.14   Inter-Node B synchronised serving HS-DSCH cell change after active set update call flow 2/3

combined radio link addition and the serving HS-DSCH cell change based on received measurement reports and/or load control algorithms (measurements may be performed in compressed mode for FDD).

First the SRNC establishes the new radio link in the target Node B for the dedicated physical channels and transmits an RRC Active Set Update message to the UE using the old radio link. The Active Set Update message includes the necessary information for the establishment of the dedicated physical channels in the added radio link, especially the PSC of the new cell to be added to the active set, but no information about the changed HS-

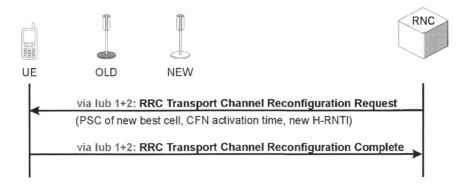

Figure 3.15   Inter-Node B synchronised serving HS-DSCH cell change after active set update call flow 3/3

PDSCH is included. When the UE has added the new radio link it returns an Active Set Update Complete message.

The SRNC will now carry on with the next step of the procedure, which is the serving HS-DSCH cell change. The target HS-DSCH cell is the newly added radio link, so far only including the dedicated physical channels. For the synchronised serving HS-DSCH cell change, both the source and the target Node Bs are first prepared for execution of the handover at the activation time.

The SRNC then sends a Transport Channel Reconfiguration message, which indicates the target HS-DSCH cell and the activation time to the UE. The message may also include a configuration of transport channel related parameters for the target HS-DSCH cell, including an indication to the transfer Iub transport bearer of MAC-d flow to the new cell. Following this procedure a status report for each RLC entity associated with the HS-DSCH should be generated.

When the UE has completed the serving HS-DSCH cell change it returns a Transport Channel Reconfiguration Complete message to the network.

Glossary

2G	second generation (synonym for GSM)
3G	third generation (synonym for UMTS)
3GPP	Third Generation Partnership Project
AAL	ATM adaptive layer
AAL2	ATM adaptation layer type 2
AAL5	ATM adaptation layer type 5
Abis	interface between BSC and BTS in GSM networks
AC	admission control; authentication centre
ACK	acknowledge
ACOM	Attach Complete (message)
AICH	acquisition indication channel
ALCAP	access link control application part
AM	acknowledge mode in RLC
AMR	adaptive multi rate
ANSI	American National Standards Institute
AP	application part
APCA	Activate PDP Context Accept (message)
APCR	Activate PDP Context Request (message)
APN	access point Name
ARFCN	absolute radio frequency channel number
ASN.1	abstract syntax notation one
ATM	asynchronous transfer mode
ATRJ	Attach Reject (message)
ATRQ	Attach Request (message)
BCC	base station colour code
BCCH	broadcast control channel
BCH	broadcast channel
BCSM	basic call state model
BER	bit error rate

BLER	block error rate
BSC	base station controller
BSIC	base station identification code
BSS	base station subsystem
BSSMAP	base station subsystem management application part
BTS	base transceiver station
CAMEL	customised application for mobile network enhanced logic
CAP	CAMEL application part
CC	call control
CC	convolutional coding (coding type)
CCCH	common control channel
CDF	continuous distribution function
CCPCH	common control physical channel
CCTrCH	coded composite transport channel
CDMA	code division multiple access
CDMA2000	3rd generation code division multiple access
CDR	call detail record
CFN	connection frame number
CGI	cell global identity
CGN	charging gateway node
c-ID	cell identity (NBAP ASN.1)
C-ID	cell identity
CID	connection identifier (AAL2)
ChCd	channelisation code
CN	core network
CORBA	common object request broker architecture
CPCH	common packet channel
CPICH	common pilot channel
CRC	cyclic redundancy check
CREF	connection refused (SCCP)
CRNC	controlling radio network controller
CS	circuit switched
CS-1/2	capability set 1/2 (INAP)
CS-iRAT-HO	circuit switched inter-RAT handover
CTCH	common traffic channel
CT/Fi	call trace/filter (application)
CTrCH	common transport channel
dB	decibel
dBm	decibel referenced to 1 milliwatt
DCCH	dedicated control channel
DCH	dedicated channel
DECT	digital enhanced cordless telephone
DL	downlink
DL BLER	downlink block error rate

DLR	destination local reference number (SCCP)
DPCH	dedicated physical channel
DPCCH	dedicated physical control channel
DPDCH	dedicated physical data channel
DRNC	drift radio network controller
DRNS	drift radio network subsystem
DSCH	downlink shared channel
DTAP	direct transfer application part
DTCH	dedicated traffic channel
Ec/N0	chip energy over noise (measured on CPICH)
EDGE	enhanced data rates for GSM evolution
E-GPRS	enhanced GPRS
ETSI	European Telecommunication Standards Institute
FACH	forward access channel
FBI	feedback information
FDD	frequency division duplex
FER	frame error rate
FP	frame protocol
FTP	file transfer protocol
Gb	GPRS interface between SGSN and GSM BSS
GERAN	GSM/EDGE radio access network
GGSN	gateway GPRS support node
Gi	interface between GGSN and external network
GMM	GPRS mobility management
GMSC	gateway MSC
Gn	interface between two GSNs
Gp	interface between two GGSNs
GPRS	general packet radio service
Gr	interface between SGSN and HLR/AuC
Gs	interface between SGSN and serving MSC/VLR
GSM	Global System for Mobile Communication
GSN	GPRS support nodes
GTP	GPRS tunnelling protocol
GTP-C	GTP control
GTP-U	GTP user
GUI	graphic user interface
H.323	an ITU-T protocol suite for video-telephony over IP
H.263	an ITU-T video codec
HARQ	hybrid automatic repeat request
HLR	home location register
HO	handover
H-RNTI	high speed downlink shared channel radio network temporary identity

HSDPA	high speed downlink packet access
HS-DSCH	high speed downlink shared channel
HSPA	high speed packet access
HS-PDSCH	high speed physical downlink shared channel
HS-SCCH	high speed shared control channel
HTTP	hyper text transfer protocol
HW	hardware
Hz	hertz, cycles per second
ID	identifier
IFHO	inter-frequency (hard) handover
IMEI	international mobile equipment identity
IMS	IP multimedia subsystem
IMSI	international mobile subscriber identity
IN	intelligent network
INAP	intelligent network application part
InFHHO	intra-frequency hard handover
InitMsg	Initiating Message
IP	Internet protocol
IP/ST	Internet protocol in ST datagram mode
IPv4	Internet protocol version 4
IPv6	Internet protocol version 6
ISCP	interference signal code power
ISDN	integrated services digital network
ISO	International Organization for Standardization
ISP	Internet service provider
ITU	International Telecommunication Union
Iu	UMTS interface between 3G-MSC/SGSN and RNC
Iub	UMTS interface between RNC and Node B
IuCS	UTRAN interface between RNC and the circuit-switched domain of the CN
IuPS	UTRAN interface between RNC and the packet-switched domain of the CN
Iur	UMTS interface between RNCs
kbps	kilobits per second
kHz	kilohertz
KPI	key performance indicator
KQI	key quality indicator
LA	location area
LAC	location area code
LAI	location area identity
LAN	local area network
LDC	long distance carrier
LI	length indicator

MAC	medium access control; message authentication code
MAP	mobile application part
Mbps	megabits per second
MCC	mobile country code
Mcps	megachips per second
Meas-ID	measurement ID (NBAP)
MEHO	mobile evaluated handover
MGW	media gateway
MHz	megahertz
MM	mobility management
MMS	multimedia messaging service
MNC	mobile network code
MOC	mobile originated call
MPEG	Moving Pictures Expert Group
MS	mobile station
MSC	mobile switching centre
MSISDN	mobile subscriber ISDN number
MSN	mobile subscriber number
MTC	mobile terminated call
N/A	not assigned
NAS	non access stratum
NBAP	Node B application part
NCC	network colour code
NE	network elements
NEM	network equipment Manufacturer
NMS	network management subsystem
Node B	UMTS base station
OMC	operation and maintenance centre
OSI	open system interconnection
O&M	operation and maintenance
PCCH	paging control channel
P-CCPCH	primary common control physical channel
PCH	paging channel
PDF	probability density function; probability distribution function
PDP	packet data protocol (e.g. PPP, IP, X.25)
PDSCH	physical downlink shared channel
PDU	packet data unit
PER	packed encoding rules (ASN.1)
PhCh	physical channel
PIP	'P' Internet protocol
PLMN	public land mobile network
PM	performance Management; performance measurement
PO	power offset

PPP	point-to-point protocol
PRACH	physical random access channel
PS	packet switched
PSC	primary scrambling code
PS-iRAT-HO	packet switched inter-RAT handover
P-TMSI	packet TMSI
QAM	quadrature amplitude modulation
QE	quality estimate
QoS	quality of service
QoE	quality of experience
QPSK	quadrature phase shift keying
R4	Release 4 of 3GPP UMTS standard
R5	Release 5 of 3GPP UMTS standard
R99	Release 1999 of 3GPP UMTS standard, sometimes called Release 3 (R3)
RA	routing area
RAB	radio access bearer
RAB-ID	radio access bearer Identity
RACH	random access channel
RAI	routing area identity
RAN	radio access network
RANAP	radio access network application part
RAT	radio access technology
RAU	routing area update
RB	radio bearer
RF	radio frequency
RL	radio link
RL-ID	radio link identity
RLC	radio link control; release complete message
RLS	radio link set
RNC	radio network controller
RNS	radio network subsystem
RNSAP	radio network subsystem application part
RNTI	radio network temporary identity
RRC	radio resource control
RSCP	received signal code power
RSSI	received signal strength indicator
RTWP	received total wideband power
RTT	round trip time
SAC	service area code
SAI	service area identity
SAP	service access point
SCCP	signalling connection control part

S-CCPCH	secondary common control physical channel
SCH	synchronisation channel
SDU	service data unit
SF	spreading factor
SFN	system frame number
SGSN	serving GPRS support node
SIB	system information block
SIM	subscriber identity module
SIR	signal-to-interference ratio
SLA	service level agreement
SLR	source local reference number (SCCP)
SM	session management
SMS	short message service
SPC	signalling point code
SRB	signalling radio bearer
SRNC	serving radio network controller
SRNS	serving radio network subsystem
SSCOP	service specific connection oriented protocol
SSDT	site selection diversity transmission
STM	synchronous transfer module
STM1	synchronous transport module – level 1
SuccOut	Successful Outcome (message)
SVC	switched virtual connection
SW	software
TB	transport block
TBS	transport block set
TC	turbo coding (coding type)
TCP	transmission control protocol
TDD	time division duplex
TDMA	time division multiple access
TD-SCDMA	time division – synchronised code division multiple access
TEID	tunnel endpoint identifier
TF	transport format
TFC	transport format combination
TFCI	transport format combination indicator
TFS	transport format set
TMF	Tele Management Forum
TMSI	temporary mobile subscriber identity
TPC	transmission power control; transmit power command
TP/IX	the 'next' Internet protocol
TR	technical report (3GPP, ETSI)
TrCH	transport channel
TRX	transceiver
TS	technical specification (3GPP, ETSI)
TTI	time transmission interval

TUBA	TCP/UDP over connectionless-mode network layer protocol (CLNP)
TV	television
Tx	transmission
uARFCN	UMTS absolute radio frequency channel number
UDP	user datagram protocol
UE	user equipment
UL	uplink
UL BLER	uplink block error rate
UL-SC	uplink scrambling code
UL-SF	uplink spreading factor
Um	GSM air interface
UM	unacknowledged mode in RLC
UMTS	Universal Mobile Telecommunication System
UP	user plane
URA	UMTS registration area
USIM	UMTS subscriber identity module
UTRA	UMTS terrestrial radio access
UTRAN	UMTS terrestrial radio access network
Uu	UMTS air interface
VCI	virtual channel identifier
VLR	visitor location register
VoIP	voice over IP
VPI	virtual path identifier
WAP	wireless application protocol
WCDMA, W-CDMA	wideband code division multiple access
WiFi	wireless fidelity
WLAN	wireless local area network
WWW	World Wide Web

References

TECHNICAL SPECIFICATIONS

3GPP Technical Specifications; **http://www.3gpp.org**

The 3GPP specifications referred to in this book are from the release 99 and release 5 set of specifications.

European Telecommunication Standards Institute; **http://www.etsi.org**

Internet Engineering Taskforce Specifications; **http://www.ietf.org**

International Telecommunication Union; **http://www.itu.int**

EXTRACT OF UMTS-RELATED SPECIFICATIONS

3GPP 23.107	Quality of Service (QoS) Concept and Architecture
3GPP 23.110	UMTS Access Stratum Services and Functions
3GPP 25.133	Requirements for Support of Radio Resource Management (FDD)
3GPP 25.211	Physical Channels and Mapping of Transport Channels onto Physical Channels (FDD)
3GPP 25.212	Technical Specification Group Radio Access Network: Multiplexing and Channel Coding (FDD)
3GPP 25.213	Technical Specification Group Radio Access Network: Spreading and Modulation (FDD)
3GPP 25.214	Technical Specification Group Radio Access Network: Physical Layer Procedures (FDD)
3GPP 25.215	Physical Layer – Measurements (FDD)
3GPP 25.301	Radio Interface Protocol Architecture
3GPP 25.308	High Speed Downlink Packet Access (HSDPA): Overall description: Stage 2
3GPP 25.321	Medium Access Control (MAC) Protocol Specification
3GPP 25.322	Radio Link Control (RLC) Protocol Specification
3GPP 25.323	Packet Data Convergence Protocol (PDCP) protocol
3GPP 25.324	Radio Interface for Broadcast/Multicast Services
3GPP 25.331	Radio Resource Control (RRC) Protocol Specification
3GPP 25.401	UTRAN Overall Description
3GPP 25.410	UTRAN Iu Interface: General Aspects and Principles
3GPP 25.411	UTRAN Iu Interface Layer 1
3GPP 25.413	UTRAN Iu Interface: RANAP Signalling
3GPP 25.420	UTRAN Iur Interface: General Aspects and Principles
3GPP 25.423	UTRAN Iur Interface RNSAP Signalling
3GPP 25.427	UTRAN Iub/Iur Interface User Plane Protocol for DCH Data Streams

3GPP 25.430	UTRAN Iub Interface: General Aspects and Principles
3GPP 25.433	UTRAN Iub Interface NBAP Signalling
3GPP 25.858	High Speed Downlink Packet Access: Physical Layer Aspects
3GPP 25.877	High Speed Downlink Packet Access: Iub/Iur Protocol Aspects
3GPP 26.101	Mandatory Speech Codec Speech Processing Functions: Adaptive Multi-Rate (AMR) Speech Codec Frame Structure
3GPP 26.091	Mandatory Speech Codec Speech Processing Functions: Adaptive Multi-Rate (AMR) Speech Codec: Error Concealment of Lost Frames
3GPP 29.061	GPRS Tunnelling Protocol (GPT) across the Gn and Gp Interface CCITT Rec. E.880 Field Data Collection and Evaluation on the Performance of Equipment, Network, and Services
3GPP 32.101	Telecommunication Management: Principles and High-Level Requirements
3GPP 32.403	Telecommunication Management; Performance Management (PM): Performance Measurements – UMTS and Combined UMTS/GSM
3GPP 34.108	Common Test Environments for User Equipment (UE): Conformance Testing
ETSI GSM 12.04	Digital Cellular Telecommunication System (Phase 2): Performance Data Measurements
IETF RFC 791	Internet Protocol
IETF RFC 768	User Datagram Protocol
IETF RFC 2225	Classical IP and ARP over ATM
IETF RFC 2460	Internet Protocol, Version 6 (IPv6) Specification
ITU-T I.361	B-ISDN ATM Layer Specification
ITU-T I.363.2	B-ISDN ATM Adaptation Layer Type 2
ITU-T I.363.5	B-ISDN ATM Adaptation Layer Type 5
ITU-T Q.711	Functional Description of the Signalling Connection Control Part
ITU-T Q.712	Definition and Function of Signalling Connection Control Part Messages
ITU-T Q.713	Signalling Connection Control Part Formats and Codes
ITU-T Q.714	Signalling Connection Control Part Procedures
ITU-T Q.715	Signalling Connection Control Part User Guide
ITU-T Q.716	Signalling Connection Control Part (SCCP) Performance
ITU-T Q.2100	B-ISDN Signalling ATM Adaptation Layer (SAAL) – Overview Description
ITU-T Q.2110	B-ISDN ATM Adaptation Layer – Service Specific Connection Oriented Protocol (SSCOP)
ITU-T Q.2150.1	B-ISDN ATM Adaptation Layer-Signalling Transport Converter for the MTP3b
ITU-T Q.2150.2	AAL Type 2 Signalling Transport Converter on SSCOP (Draft)
ITU-T Q.2630.1	AAL type 2 Signalling Protocol (Capability Set 1)

LITERATURE

Holma, Toskala, *WCDMA for UMTS*, John Wiley & Sons, Ltd, UK, 2001.

Kaaranen, Naghian, Laitinen, Ahtiainen, Niemi, *UMTS Networks*, John Wiley & Sons, Ltd, UK, 2001.

Kreher, Ruedebusch, *UMTS Signaling*, John Wiley & Sons, Ltd, UK, 2005.

Laiho, Wacker, Novosad, *Radio Network Planning and Optimisation for UMTS*, John Wiley & Sons, Ltd, UK 2002

OTHER WWW SOURCES

Vallent: Leader in Service Assurance – www.watchmark.com

NIST/SEMATECH e-Handbook of Statistical Methods – http://www.itl.nist.gov/div898/handbook/, January 2006.

Rysavy Research – Wireless Technology: Assessment and Integration – www.rysavy.com

3G Americas: Unifying the Americas through Wireless Technology – www.3gamericas.org

Index

Printed and bound in the UK by
CPI Antony Rowe, Eastbourne

Printed and bound by CPI Group (UK) Ltd, Croydon, CR0 4YY

16/04/2025

14658476-0001